One Earth, One Chance

A Practical Guide to Good Environmental Stewardship

Sami Elkhayri

Contents

Introduction

Your Role as a Steward

What it means to be a good steward of our shared home

Being a good steward of our planet means recognizing that we don't own nature. We care for it. And, in return, it cares for us. Stewardship is the simple yet powerful idea that each of us has a role in protecting and wisely using Earth's resources so they remain healthy and abundant for everyone, not just for us today, but for generations to come. It's not about being perfect; it's about making choices, big and small, that help restore balance where we live, work, and play. From how we shop and travel to how we vote and speak up for the voiceless—whether that's wildlife, forests, or future generations—stewardship means acting as guardians of our shared home so that life in all of its forms can thrive.

Why Your Actions Matter

It's easy to feel that one person's choices won't make a difference when the problems seem so huge—melting ice caps, burning forests, plastic-choked oceans. But the truth is that real change always starts with ordinary people taking small, steady actions that add up over time. When we conserve water, use less energy, choose sustainable products, and speak up for the natural world, we're sending ripples through our families, communities, and the wider world. As a steward, you're not just changing your habits—you're setting an example and creating a culture of care that inspires others to do the same.

Stewardship Is Local—And Global

Stewardship starts close to home: your neighbourhood park, your local river, or the soil in your backyard. But in our connected world, our choices reach far beyond our doorstep. The coffee we drink, the clothes we buy, and the forests we protect all link us to people and places around the globe. That's why becoming a good steward means adopting a two-pronged strategy: protecting what's nearby and caring about what happens oceans away. By taking mindful actions where we live, and supporting fair, sustainable practises everywhere, we help weave a safety net for our entire planet.

Moving Forward

As you turn the pages of this book, you'll discover that becoming a good steward doesn't require perfection or expertise—only a willingness to learn, reflect, and act. You'll meet real people who are making a difference, explore the planet's most urgent challenges, and find practical, doable steps that fit your everyday life.

This isn't about doing everything. It's about doing something and knowing that, when enough of us do, it adds up to real, lasting change. You are not alone in this. You are part of a growing movement of people who care deeply about the Earth and are ready to protect it. Let's begin this journey together.

Part I

Understanding What We're Protecting

Chapter 1
Our Shared Planet

Nature as Our Life-Support System: Air, Water, Soil, Food

When we think of life-support systems, we might picture something high-tech—astronauts floating in space, connected to tubes and oxygen tanks. But, here on Earth, our true life-support system is nature itself. Every breath we take, every glass of clean water we drink, every bite of food we eat depends on healthy, living ecosystems that keep our planet in balance.

Take the air, for example. Forests, wetlands, grasslands, and even ocean plankton all help clean our air by absorbing carbon dioxide and releasing oxygen. A single mature tree can produce enough oxygen in a year to support two people. When we protect forests and plant more trees, we're literally investing in the air we breathe.

Water is just as miraculous—and just as vulnerable. Rivers, lakes, wetlands, and underground aquifers store, filter, and deliver fresh water to our homes, farms, and cities. Wetlands alone can filter out pollutants and act as sponges during floods and droughts. But when we drain wetlands, pave over watersheds, or dump waste into rivers, we break this natural purification cycle—and we end up paying the price through water shortages, costly treatment plants, or polluted drinking water.

Soil, too, is a quiet hero. Healthy soil is alive with millions of organisms that break down organic matter, recycle nutrients, and hold water like a sponge. This living soil grows our food, nourishes our forests, and supports the plants and animals that make up local ecosystems. But when soil is overused or contaminated with chemicals, it loses its life—and our food systems become fragile.

And food? It connects all these systems together. What we eat, how it's grown, and how far it travels all affect nature's balance. Sustainable farming that works with nature—protecting pollinators, planting diverse crops, conserving water—keeps soil healthy and ecosystems thriving. When we make thoughtful choices about what's on our plate, we're not just feeding ourselves—we're helping maintain the planet's life-support systems for everyone.

This is the heart of stewardship: recognizing that nature does the hard work of sustaining us every single day. In return, our job is to care for these systems, keep them healthy, and restore them when they're damaged. The truth is, when nature thrives, we thrive too.

Mini-Reflection: Your Invisible Connection

Next time you step outside, pause and notice the invisible threads that connect you to the natural world. Take a deep breath—that's a gift from trees, plants, and plankton. Pour a glass of water—it likely fell as rain, trickled through soil, or flowed through wetlands that filtered it for you. Eat a piece of fruit—soil, pollinators, and healthy ecosystems made it possible.

These life-giving connections are easy to overlook but impossible to live without. By remembering them, we remember why our everyday choices matter.

Action Tip: One Small Step Today

Action Tip: Pick one habit, this week, that shows respect for your life-support system.

- Turn off the tap while brushing your teeth to save water.
- Plant a native flower or tree to clean the air and help pollinators.
- Compost food scraps to feed the soil.
- Choose a meal with local, seasonal ingredients.

Stewardship starts right where you are—one mindful choice at a time.

What Conservation Really Means—Not Just Protection, But Wise Use

When many people hear the word *conservation*, they think of setting land aside, building fences, and keeping humans out so nature can heal. While protected areas, like national parks and wildlife reserves, are vital, conservation is so much bigger than simply drawing a line on a map. True conservation is about wisely using—not overusing—the Earth's resources so they stay healthy and abundant for everyone, now and in the future.

Think of it like this: conservation asks, *How can we live with nature, not just alongside it?* Indigenous communities around the world have practised this balance for thousands of years. Farmers rotate crops to protect soil, fishers follow seasonal patterns to avoid depleting fish stocks, and forest communities harvest plants or timber in ways that allow trees to regrow. These practises come from a simple truth: when we take too much or take too quickly, nature can't keep up—and everyone loses.

Conservation means asking better questions before we act:

- How much is enough for us—and how much does nature need to keep working?
- Can we take what we need without destroying what we'll need tomorrow?
- Who depends on this forest, river, or ocean besides us—birds, fish, future generations?

Wise use also means looking at waste. It's not conservation if we take and toss mindlessly. When we waste food, we waste the water, soil, energy, and human labour that went into growing it. When we throw away products that could be reused or recycled, we put unnecessary strain on ecosystems that provide raw materials.

Being a good steward means embracing conservation as an everyday practise, not just a distant policy. It's switching to renewable energy instead of burning up fossil fuels that we can't replace. It's choosing sustainably harvested wood, fair-trade coffee, or local produce that doesn't drain faraway lands. It's saying *no* to overconsumption and *yes* to living well within nature's limits.

The goal isn't to shut ourselves out of nature—it's to remind ourselves that we're part of it. When we use resources wisely, we show respect for the web of life that supports us all. Conservation isn't just about protecting wild places—it's about protecting our future by living like we know we belong here.

Mini-Reflection: Enough for All

Take a moment to think about something you use every day—like water, wood, or seafood. Who else relies on that same resource? Birds nesting in trees, fish feeding local communities, and future generations yet to be born. Conservation is about seeing these invisible connections and asking, *How can I take only what I truly need so there's enough for everyone—now and later?* It's a powerful mindset shift that turns every choice into an act of care.

Action Tip: Start Small, Use Wisely

Action Tip: This week, pick one everyday resource and use it a little more wisely.

- Take shorter showers to save water for rivers and wetlands.
- Eat seafood from sustainable sources to protect ocean life.
- Choose recycled or reclaimed wood products.
- Reduce food waste by planning meals mindfully.

Small acts of wise use add up—they show that you're part of the solution, not the problem.

Stewardship in Action: Māori Guardianship ('Kaitiakitanga') in New Zealand[1]

When we talk about stewardship, we can learn so much from communities that have practised it for generations. One powerful example is *kaitiakitanga*—the Māori concept of guardianship in *Aotearoa* (the indigenous name of New Zealand). For *Māori*, the Indigenous people of New Zealand, caring for the natural world isn't a choice or a trend—it's a deep responsibility that comes from their *whakapapa* (ancestral connections) to the land, sea, and all living things.

Kaitiakitanga comes from the word *kaitiaki*, which means caretaker, guardian, or steward. But it's more than just a job—it's a living relationship. The idea is that humans are part of the natural world, not separate from it, and that nature has its own life force (*mauri*) that must be protected. Māori believe that if the *mauri* of a river, forest, or coastline is healthy, then the people and communities connected to it will be healthy too.

You can see *kaitiakitanga* in action across New Zealand today. Māori *iwi* (tribes) and *hapū* (sub-tribes) help manage forests, rivers, and marine areas using traditional knowledge alongside modern science. Some communities have worked to restore polluted rivers, reintroduce native species, and limit fishing in certain areas so fish populations can recover. Many of these projects are legally supported, recognizing *Māori* rights as treaty partners and custodians of their ancestral lands.

For example, the *Whanganui* River was granted legal personhood in 2017—meaning that it's recognized as a living entity with its rights. Local *Māori* act as *kaitiaki* for the river, ensuring its well-being is upheld in decisions about land use, pollution, and resource extraction. This way of thinking shifts the focus from ownership and exploitation to care and respect.

Kaitiakitanga reminds us that faithful stewardship is not just about protecting nature from people—it's about people protecting nature *with* respect, wisdom, and humility. It shows us that when we see ourselves as guardians, not owners, we make choices that sustain life for generations to come.

Reflection: What Can We Learn?

Ask yourself: How can you bring the spirit of *kaitiakitanga* into your own life? What nearby place—a park, stream, or garden—could you care for as a living being, not just a resource? When you see yourself as a guardian, your everyday actions become a promise to future generations.

Stewardship in Action: The Bishnoi Community in India[2]

Long before modern conservation movements, the Bishnoi community of Rajasthan, India, practised a way of life rooted in harmony with nature. The Bishnoi follow 29 principles (the word *Bishnoi* comes from *bis* = twenty, *noi* = nine) laid out by their 15th-century founder, Guru Jambheshwar. These include protecting trees and animals, as well as living in a way that avoids harming any living being.

One of the most famous Bishnoi stories is the sacrifice of the *Amrita Devi* and her village in 1730. When royal soldiers came to cut down sacred *khejri* trees for a new palace, the villagers hugged the trees to protect them—giving their lives rather than seeing them destroyed. This courageous act inspired India's *Chipko* Movement centuries later, where villagers literally embraced trees to stop deforestation.

Today, the Bishnoi still protect wildlife in their desert communities, sharing water with antelopes, feeding birds, and enforcing community rules against poaching. Their deep respect for nature shows us that stewardship can be a daily practise woven into culture, faith, and identity.

Stewardship in Action: Community Forests in Nepal[3]

In Nepal, local communities have helped transform deforested hillsides into lush green forests through community forestry. After years of deforestation and poor management, the government began giving villagers legal rights to manage their local forests themselves.

Villagers formed forest-user groups to decide how much wood to cut, where to plant trees, and how to share benefits fairly. They use local knowledge and modern science to monitor forest health, prevent illegal logging, and keep wildlife habitats thriving.

This approach has been so successful that community-managed forests now make up about one-third of Nepal's forest area. Communities earn income from forest products, while protecting watersheds and biodiversity. Nepal's story shows that when local people are empowered to be stewards, both nature and livelihoods flourish.

Stewardship in Action: Traditional Fishing Rights in the Pacific[4]

Across the Pacific Islands, coastal communities have long practised stewardship through traditional marine tenure. Villages manage their reefs and fisheries using customs passed down for generations—such as rotating fishing grounds, setting seasonal closures, or declaring certain areas as sacred and off-limits.

In places like Fiji, Samoa, and the Solomon Islands, communities today blend these ancestral rules with modern conservation science. They create locally managed marine areas (LMMAs) to rebuild fish stocks and protect coral reefs. Studies show that these areas often have healthier fish populations than government-managed reserves.

This reminds us that stewardship isn't only about saying *no* to using resources—it's about using them wisely, respecting nature's limits, and passing on abundance to the next generations.

Stewardship in Action: Regenerative Farmers in North America[5]

All around North America, a growing number of farmers are practising regenerative agriculture—farming that works with nature to restore soil health, increase biodiversity, and store carbon in the ground.

Regenerative farmers use techniques like cover cropping, rotating livestock, planting diverse crops, and minimizing tilling the soil. These practises reduce erosion, improve water retention, and make farms more resilient to droughts and floods.

By seeing themselves as stewards of the land, these farmers produce healthy food while helping tackle climate change. They show us that conservation isn't only about setting land aside—it's about how we work the land, too.

Reflection: What Can We Learn?

Ask yourself: What lessons from these stories can you bring into your own life? What tradition, practise, or local knowledge could inspire you to be a better steward of your place?

Your Stewardship Steps: Getting Started

Reflect

Pause for a moment and think about the nature around you. Where do you see signs of nature in your daily life—trees along a street, a patch of wildflowers, a local park, birds outside your window? What would your neighbourhood feel like without these living things? Try to picture how your day would change if these small parts of nature disappeared. This simple reflection helps you realize you're not separate from nature—you're part of it.

Learn

Take 10 minutes to look up who is already caring for your local environment. Is there a conservation group protecting a nearby park or river? Are indigenous communities involved in stewardship where you live? Read a short article or visit an organization's website to see what challenges they're tackling and how people like you can help. Knowing who's already doing the work reminds you you're not alone—you're part of a community of stewards.

Act

Do one small thing this week that directly cares for your local piece of the planet. Pick up litter in your neighbourhood park, plant a native flower or tree, or talk to a friend or neighbour about something you learned in this chapter. Tiny actions remind us that we're not powerless—we're caretakers of the places we love.

Practical Takeaways:

Map your local ecosystem—parks, rivers, or green spaces you rely on.

Start a simple journal: how your daily choices affect these.

Stewardship Journal

What did I learn?

_____

_____

_____

_____

_____

_____

What small action will I try this week?

_____

_____

_____

_____

_____

_____

With whom will I share what I have learned?

_____

_____

_____

_____

_____

Part II

The Big Issues, the Everyday Solutions

Chapter 2

Earth's Life-Support

A Plain-Language Snapshot of the Problem

We often take our planet's life-support systems for granted because they work quietly in the background. But the truth is, the basic natural systems we depend on—clean air, fresh water, healthy soil, a stable climate—are under enormous pressure.

Every year, we're using up Earth's natural resources faster than they can regenerate. We pump greenhouse gases into the atmosphere at record rates, warming the climate and creating extreme weather that disrupts communities and crops. We clear forests for timber, agriculture, and development, wiping out the trees that filter our air and store carbon. We pollute rivers and oceans with waste, chemicals, and plastics, threatening the water we drink and the food we eat.

Soil, too, is suffering. Industrial farming methods, pesticides, and overuse strip soil of its nutrients and its living organisms, making it harder to grow healthy food. In some regions, topsoil is eroding so quickly that farmland may become barren in a generation if we don't change course.

Meanwhile, our appetite for resources keeps growing. The average person today uses more energy, water, and materials than ever before. Our *take, make, waste* habits add up to a massive drain on ecosystems and push nature's limits to the breaking point.

The good news is, nature is incredibly resilient—but only if we give it a chance. By understanding how we rely on these life-support systems and what's putting them at risk, we can make more intelligent choices. Conservation isn't just about saving nature—it's about saving the systems that keep all of us alive, now and for generations to come.

Mini-Reflection: A Shared Responsibility

Pause and think about the last thing you threw away, the last time you left a light on, or the last drive you made alone in the car. None of us is perfect—but each small habit adds up when billions of people do the same thing. Now flip the thought: what if billions of us made one wiser choice each day instead? Remember: you are not powerless. The same everyday actions that contribute to the problem can become part of the solution.

Action Tip: One Step Toward Balance

Action Tip: Pick one life-support system—air, water, soil, or climate—and protect it in a simple way today:

- Air: Walk, bike, or take public transit instead of driving once this week.
- Water: Fix a leak or use a rain barrel to save fresh water.
- Soil: Compost your food scraps instead of tossing them.

- Climate: Switch off lights and unplug devices you're not using. Small steps help ease the pressure on the systems that keep us all alive.

Where Your Carbon Footprint Comes From[6]

When you hear about climate change, you'll often hear the term *carbon footprint*. But what does that really mean? In plain language, your carbon footprint is the total amount of greenhouse gases—primarily carbon dioxide (CO_2), but also methane (CH_4) and others—released into the atmosphere because of your everyday activities.

For most people, the biggest piece of their footprint comes from energy—mainly burning fossil fuels like coal, oil, and natural gas for electricity, heating, and transportation. For example, driving a gasoline-powered car or flying in a plane burns fuel that releases carbon dioxide into the air. Heating your home with oil or natural gas does the same. Even the electricity that powers your lights, appliances, and devices has a footprint, depending on how your local grid produces that energy.

Food is another significant contributor to our carbon footprint. Producing food uses energy to run tractors, process crops, and transport goods around the world. Raising livestock like cattle generates methane, a powerful greenhouse gas. Food waste makes it worse: when we throw food away, all the resources it took to grow, ship, and store that food are wasted too—plus rotting food in landfills releases methane.

The stuff we buy—clothes, electronics, furniture—also adds up. Everything we use requires energy and resources to make, package, and ship. The more disposable our habits, the bigger our footprint grows. Even our everyday digital life, such as streaming videos or storing files in the cloud, relies on electricity from massive data centres.

It might feel overwhelming, but the good news is there are simple ways to shrink your carbon footprint: drive less, fly less, eat more plant-based meals, waste less food, buy fewer things, and choose items that last. Small actions at home, at work, and in your community all add up when millions of us do them together.

Understanding where your carbon footprint comes from is the first step to taking responsibility for it—and it reminds us that our daily choices can protect the Earth's life-support systems for everyone.

Mini-Reflection: What's One Big Source for You?

Think about your day: where do you think the biggest chunk of your carbon footprint comes from—driving, flying, food, electricity, or stuff you buy? Pick one area to explore more closely this week. Awareness is the first step toward change.

Action Tip: Quick Carbon Cut

Action Tip: Try one small switch:

- Walk, bike, or take public transit for one trip this week instead of driving.
- Plan a plant-based meal.
- Unplug devices you rarely use.
- Hold off buying something new—repair or borrow instead.

Little by little, you'll lighten your footprint—and help the planet breathe easier.

Carbon Footprint 101: A Quick Self-Check

Use this simple checklist to see which parts of your life add the most to your carbon footprint. You don't need exact numbers—just circle or jot down what applies to you. Awareness is your first step toward action!

Transportation

How do you usually get around?

◊   Drive alone most days
◊   Carpool sometimes
◊   Use public transit often
◊   Walk or bike for some trips

Do you fly often for work or vacation?

◊   0-1 flights/year
◊   2-5 flights/year
◊   More than 5 flights/year

Energy Use at Home

What powers your heating and cooling?

◊   Oil / Gas
◊   Electricity
◊   Solar / Renewable

Do you switch off lights and devices when not needed?

◊   Always
◊   Sometimes
◊   Rarely

Food Choices

How often do you eat red meat?

◊   Every day
◊   A few times a week
◊   Rarely / never

Do you waste much food?

◊   Hardly any
◊   Some
◊   A lot

Stuff You Buy

How often do you buy new clothes, gadgets, or household goods?

◊ Very often
◊ Sometimes
◊ Rarely—I repair or reuse

What's My Biggest Chunk?

Take a look at your answers. What stands out as your biggest source of carbon emissions? Mark it with a ⊠.

One Thing I Can Try:

Write down one thing you'll do this week to shrink that source:

"This week I will _____."

Remember: small actions add up—especially when millions of us do them together!

Tip: For a deeper dive, try an online carbon footprint calculator like WWF, Global Footprint Network, or your local government's sustainability site.

Case Studies

Case Study: Denmark's Transition to Wind Energy[7]

When people talk about clean energy success stories, Denmark is one of the best examples in the world. This small Scandinavian country has shown what's possible when a nation commits to using nature's life-support systems wisely—in this case, the power of the wind.

Back in the 1970s, Denmark relied heavily on imported oil for its energy. But when the oil crisis hit, people realized they needed a more secure and sustainable way to power their homes and economy. The Danish government, local communities, and innovative companies came together to try something bold: invest in wind energy.

They didn't just build giant turbines overnight. Communities formed local co-ops so that people living near wind farms could actually own a share of the turbines—and the profits. This built public trust and pride in renewable energy. Over time, Denmark also supported research and innovation, helping Danish companies become leaders in wind turbine technology around the world.

Today, Denmark gets about half of its electricity from wind. On particularly windy days, it produces more electricity than the whole country needs. Extra energy can be shared with neighbouring countries, showing how renewable energy can strengthen energy security for everyone.

Denmark's story proves that transitioning to clean energy doesn't mean going back to the Stone Age—it means tapping into the life-giving systems that nature offers freely. By harnessing the power of the wind, Denmark reduces greenhouse gas emissions, enhances air quality, and generates local jobs and community wealth.

This success didn't happen overnight—it took decades of vision, policies, community support, and technological progress. It shows what's possible when people choose to work with nature, not against it. Every gust of wind that spins a Danish turbine is a reminder that a clean energy future is not only possible—it's already here.

Case Study: Germany's Solar Boom[8]

Germany is famous for having a lot of cloudy days—yet it's also one of the world's solar energy leaders. How did that happen?

In the early 2000s, Germany introduced bold feed-in tariffs that guaranteed fair payments for households and businesses that installed solar panels and fed electricity back into the grid. Suddenly, everyone, not just big companies, could become part of the energy transition.

Tens of thousands of rooftops, barns, and warehouses became mini power plants. Farmers earned extra income, while villages and cities invested in community solar farms. This people-powered approach helped Germany produce a significant portion of its electricity from the sun—while creating local jobs and lowering emissions.

The lesson? Renewable energy works best when everyday people can take part. Solar panels don't just cut pollution—they put power (literally) back in our hands.

Case Study: Iceland's Geothermal Energy[9]

Iceland sits on a hot spot where Earth's crust is alive with geothermal energy. Instead of letting this natural resource go to waste, Iceland has become a world leader in using geothermal heat to warm homes, greenhouses, and even sidewalks!

About 90% of Icelandic homes use geothermal heating—piping naturally hot water from underground straight into radiators and taps. This has dramatically reduced the need for fossil fuels. The country also taps hydropower from rivers, making nearly all of its electricity renewable.

Iceland's example shows how smart stewardship means working with what nature offers locally. Every region has different resources—wind, sun, water, and underground heat. The key is to respect and harness them sustainably, without exhausting or damaging them.

Case Study: Costa Rica's Renewable Mix[10]

Costa Rica is a small Central American nation with a big commitment: run on nearly 100% renewable electricity. For years, it's done precisely that—drawing energy from hydropower, geothermal plants, wind farms, and some solar.

The country's abundant rivers power most of its dams. Its volcanoes supply geothermal heat. And local communities support these systems because they know clean energy protects Costa Rica's rich rainforests, rivers, and wildlife. The same natural wonders draw millions of tourists each year.

Costa Rica's story reminds us that caring for nature and building a strong economy can go hand in hand. Stewardship doesn't hold us back—it moves us forward.

Case Study: Indigenous Microgrids in Remote Communities[11]

In many remote parts of the world, communities are turning away from polluting diesel generators and building small, local microgrids powered by renewables instead.

For example, some Indigenous communities in Alaska and northern Canada now use wind turbines and solar panels to keep the lights on. In the past, these communities paid huge costs to fly in barrels of diesel fuel. Now, local clean energy means fewer emissions, lower costs, and greater energy independence.

These stories show that renewable energy isn't only for big cities—it can empower even the smallest communities to live more sustainably, on their own terms.

Reflection: What Can We Learn?

Denmark's journey shows that big change happens when governments, communities, and businesses pull together for the common good. Where could your community tap into local clean energy sources? How could you support policies, projects, or companies that bring renewable energy to more people?

The success stories remind us that there's no one-size-fits-all solution. Every community has unique natural resources—and people with ideas, skills, and passion to use them wisely. What local resource could your community tap into? Who could you support or partner with to make that happen?

Your Stewardship Steps

Reflect

Take a few quiet minutes to think about how you use the Earth's life-support systems every day. The air you breathe, the water you drink, the food you eat, the forests that clean the air, the oceans that regulate climate—these systems work silently for you. Make a quick *Daily Nature Connection Map*. Where does your water come from? Who filters your air? Which parts of your life depend on healthy soil and diverse species? This awareness is the first step to protecting what protects you.

Learn

Find out what resource issues affect your region most. Is water scarcity an issue? Overfishing? Deforestation? Look up your city or region's environmental report or check a trusted conservation website. Even just learning where your tap water comes from or where your electricity is generated can help you understand the link between your daily life and Earth's natural systems.

Act

Choose one small habit that helps conserve the Earth's basic life systems—air, water, soil, biodiversity. For example, fix a dripping tap to save water, unplug devices you're not using, or shop at a local farmers' market instead of buying out-of-season produce shipped from far away. Small shifts done daily really do protect these vital systems.

Practical Takeaways:

- Cut energy waste at home.
- Shift to renewables if possible.
- Reduce car use, fly less, and offset when you do.

Stewardship Journal

What did I learn?

_____

_____

_____

_____

_____

_____

What small action will I try this week?

_____

_____

_____

_____

_____

_____

With whom will I share what I have learned?

_____

_____

_____

_____

_____

_____

Chapter 3

Protecting Biodiversity

Why Species Matter, From Pollinators to Apex Predators

Every living thing on Earth—from the tiniest bee to the most enormous whale—plays a part in keeping nature's systems balanced and healthy. Sometimes it's easy to forget how connected we really are to these species until they're gone or struggling.

Take pollinators, for example. Bees, butterflies, bats, and birds help plants reproduce by carrying pollen from flower to flower. About one out of every three bites of food we eat depends on pollinators—from apples and almonds to chocolate and coffee. Without them, entire food chains would unravel, and we'd face huge drops in crop yields.

On the other end of the scale are apex predators—animals like wolves, big cats, sharks, or orcas. These species help regulate populations of prey and keep ecosystems in balance. For example, when wolves were reintroduced to Yellowstone National Park in the US, they controlled deer and elk numbers, which allowed over-browsed plants and trees to recover. That, in turn, brought back songbirds, beavers, and more. Scientists call this a *trophic cascade*—a single top predator can shape the entire landscape.

Even the less charismatic species matter. Soil microbes, fungi, insects, and plankton do the hidden work of recycling nutrients, storing carbon, and producing oxygen. Losing these creatures can have ripple effects we might not even see until it's too late.

When any link in the web of life weakens, the whole system becomes more fragile—and so does our own well-being. That's why protecting biodiversity isn't just about saving cute animals or rare plants; it's about protecting the life-support systems that keep our air breathable, our water clean, and our food on the table.

Each species has its place and purpose. As stewards, our job is to make sure they can thrive—because when nature's web is strong, we are strong too.

Reflection: A Living Connection

Think about the plants and animals around you—from the bees in your garden to the birds overhead. Which ones do you depend on without even realizing it? Which ones could you protect, plant for, or learn more about?

Action Tip: Small Steps for Species

Action Tip:
- Plant native flowers to feed pollinators.
- Avoid pesticides that harm bees and insects.
- Support conservation groups that protect threatened species in your region.
- Talk with family and friends about why every creature—big or small—matters.

How Species Extinctions Affect Us

When we hear about an animal going extinct, it can feel like a distant tragedy. But every species loss is like pulling a thread from a living web. Eventually, the whole system can unravel, including the parts we depend on.

When we lose pollinators like bees, crops suffer. When we lose fish species, local communities lose jobs, food security, and cultural traditions. When forests lose key animals that spread seeds or keep pests in check, the entire forest can decline. The impacts aren't just ecological—they're economic, cultural, and deeply personal.

Scientists warn that Earth is now facing a wave of extinctions at a rate much faster than natural cycles—driven by habitat destruction, pollution, climate change, and overexploitation. This loss of biodiversity makes our food systems less resilient, our water and air quality worse, and our ability to adapt to climate change weaker.

Protecting species is really about protecting ourselves. When we save a habitat for an endangered animal, we also protect clean water, healthy soil, and carbon storage for future generations. It's all connected.

Reflection: Look Closer

Look up a local animal or plant that's endangered in your region. Why is it disappearing? What does it tell you about the health of your local environment?

Action Tip: Be a Voice for Species

Action Tip: Support local or global efforts that protect habitats and endangered wildlife. Choose products that don't contribute to deforestation or habitat loss—like certified sustainable paper, wood, or seafood. Speak up for conservation policies that matter.

Stewardship in Action: Community Biodiversity Gardens[12]

All over the world, communities are turning empty lots, yards, and school grounds into biodiversity gardens that give local species a fighting chance.

For example, in some North American cities, volunteers plant native wildflowers and shrubs to create safe places for pollinators, butterflies, and birds. Schools and community groups build "bee hotels," plant milkweed for monarchs, or remove invasive plants to help native species recover.

These local projects show that you don't need vast wilderness to make a difference. Even a small patch of native plants can become an oasis for struggling species—and a living classroom for people to learn about their local environment.

When communities work together to restore habitat, they're strengthening the web of life one garden, yard, or green space at a time.

Reflection: Your Backyard Matters

Whether you have a yard, a balcony, or a shared green space, you can help bring nature back. What could you plant or protect this year?

Action Tip: Grow a Little Wild

Action Tip:

- Plant native flowers, shrubs, or trees that support local birds and insects.
- Avoid pesticides and herbicides that harm beneficial creatures.
- Leave some "messy" areas for pollinators, like fallen leaves or logs.
- Join or support local habitat restoration groups.

Stewardship in Action: Sea Turtle Protection[13]

Along coasts worldwide, sea turtles have survived for millions of years—but today they face huge threats, from plastic pollution and poaching to habitat loss. In places like Costa Rica, local communities protect nesting beaches by patrolling for poachers, moving vulnerable nests to safe hatcheries, and educating visitors to keep beaches dark (free from artificial lighting) and clean.

Thanks to these efforts, more turtle hatchlings survive to swim the seas—a hopeful sign that, when people take stewardship seriously, we can reverse decline and give species a fighting chance.

Reflection: Long-Term View

Sea turtles remind us that stewardship is about thinking far ahead. What choices could you make today that will help protect life not just now, but decades into the future?

Action Tip: Ocean-Smart Habits

Action Tip:
- Reduce single-use plastics that can end up in the ocean.
- Support sustainable seafood choices that don't harm marine habitats.
- Be a responsible traveller: never disturb nesting wildlife or leave trash behind.
- Support groups protecting critical habitats and species.

Case Study: Community-Led Wildlife Corridors in India[14]

In India, people and wildlife often live side by side—but as villages, roads, and farms expand, wild animals can get cut off from the habitats they need to survive. For species like elephants, tigers, and leopards, safe passage between forests is essential. Without connected landscapes, animals become trapped in isolated patches, which can lead to inbreeding, starvation, or conflict with people.

One inspiring solution has come from communities themselves: creating and protecting *wildlife corridors*—strips of land that connect fragmented habitats so animals can move safely. In the Nilgiri Biosphere Reserve in southern India, for example, local communities have worked with conservation groups to secure vital corridors for Asian elephants, whose ancient migration paths often cross roads and villages.

Rather than seeing wildlife as a nuisance, many villagers now see the benefits of coexisting. Community members help map elephant routes, remove obstacles like fences, and plant native trees to restore degraded patches. Some farmers get support to protect crops from elephant raids, using early warning systems and watch groups to reduce clashes.

This kind of community stewardship has huge ripple effects. It keeps animal populations healthy by allowing them to find mates and new feeding grounds. It also reduces human-wildlife conflict, which protects lives, livelihoods, and cultural connections to nature.

India's wildlife corridor projects show us that protecting biodiversity isn't just about setting aside big parks—it's about linking them, one piece at a time, with the help and leadership of local people. When communities have a voice and a stake, conservation works for both people and wildlife.

Reflection: Living Side by Side

How do people and wildlife cross paths where you live—birds, deer, coyotes, or other creatures? How could your community make more room for wildlife to move safely?

Action Tip: Support Local Connections

Action Tip:
- Keep hedgerows, tree lines, and green spaces intact to help animals travel.
- Volunteer with groups that restore habitat or plant native trees.
- Support policies that protect wildlife crossings and green corridors.
- Share what you learn—coexistence works best when everyone is on board.

Case Study: Urban Wildlife Corridors in Los Angeles, USA[15]

In the sprawling city of Los Angeles, traffic isn't just tough on people—it's a major barrier for wildlife too. For years, mountain lions in the Santa Monica Mountains have been trapped between highways, leading to inbreeding, vehicle collisions, and shrinking territory.

That's why local residents, conservationists, schools, and scientists joined forces to support the *Wallis Annenberg Wildlife Crossing*—a vegetated bridge over a major highway (US-101). Once completed, it will be the largest wildlife overpass in the world, giving animals like bobcats, coyotes, deer, and mountain lions a safe way to roam, breed, and survive.

Community support was key. People raised funds, shared the story, and got kids involved through art, science fairs, and classroom visits. They showed that, even in a big city, people care deeply about giving wildlife room to move.

This project proves that coexistence isn't just possible—it's popular. When nature is given a place in our urban lives, everyone benefits.

Reflection: Room to Roam

Cities don't have to be concrete jungles. Where could wildlife move more freely in your area—through a park, along a river, or across a road?

Action Tip: Green Your Neighbourhood

Action Tip:
- Support green infrastructure like wildlife bridges, native plant corridors, or pollinator pathways.
- Talk to city planners or local officials about including habitat in development plans.
- Plant native shrubs or trees in yards, balconies, or public spaces to create stepping stones for animals.

Case Study: Pollinator Pathways in the United Kingdom[16]

Across the UK, residents are stitching together a patchwork of pollinator-friendly gardens, meadows, and roadside wildflower strips to help bees, butterflies, and insects find food and shelter.

One example is the *B-Lines Project*, a national network of *insect highways* developed by the organization *Buglife*. It connects isolated habitats by encouraging communities, councils, and landowners to plant native wildflowers, reduce pesticide use, and manage grasslands in pollinator-friendly ways.

From schools and farms to roundabouts and backyards, people are transforming unused or neglected spaces into vibrant, blooming corridors. These routes help pollinators thrive—and, in turn, support the crops and ecosystems we rely on.

Even a single small garden or green roof becomes part of a larger living network. It's a reminder that biodiversity isn't just protected in faraway forests—it starts on your doorstep.

Reflection: Small Spaces, Big Impact

What small patch of land—a garden bed, balcony box, or sidewalk edge—could you help turn into a mini-wildlife refuge?

Action Tip: Create a Pollinator Stopover

Action Tip:
- Plant native, pesticide-free flowers that bloom at different times of year.
- Leave a shallow dish of water with pebbles for insects to drink from.
- Avoid over-trimming wild areas—"messy" spaces can be safe havens for bees and butterflies.

Case Study: Indigenous-Led Wildlife Corridors in Canada[17]

In western Canada, Indigenous communities are leading the way in restoring wildlife corridors across traditional territories. One powerful example is the *Y2Y (Yellowstone to Yukon) Conservation Initiative*, which spans two countries and brings together over 450 partners—including First Nations, scientists, governments, and communities—to protect a continuous stretch of habitat for wildlife.

From grizzly bears and caribou to fish and birds, these animals need large, connected landscapes to survive. Indigenous knowledge and stewardship are at the heart of this project. In some areas, communities are working to reconnect fragmented rivers and forests, remove barriers like fences, and support sustainable hunting, trapping, and tourism.

These projects also restore cultural ties to the land and create jobs in conservation, mapping, and ecological monitoring.

Y2Y shows that, when Indigenous voices lead, conservation becomes more holistic—grounded in respect, reciprocity, and generations of deep connection to place.

Reflection: Respecting Indigenous Stewardship

Who are the original caretakers of the land where you live? How can you support or learn from their conservation efforts?

Action Tip: Walk with Respect

Action Tip:
- Acknowledge Indigenous lands when visiting parks or natural areas.
- Support Indigenous-led conservation initiatives and land-back movements.
- Learn about local treaties, histories, and traditional ecological knowledge in your region.

Your Stewardship Steps

Reflect

Think about your personal carbon footprint. What do you do every day that uses the most energy—driving, heating or cooling your home, or flying for work or vacation? Imagine what one day in your life would look like if you made one small switch, like biking instead of driving, or eating more plant-based meals. How would it feel? Seeing your habits clearly helps you find simple places to start reducing emissions.

Learn

Spend a few minutes finding out if your city, town, or country has a climate action plan. Many places now have clear goals to cut emissions, shift to renewable energy, or build climate resilience. Knowing what's already planned helps you see where you can support these efforts—and hold leaders accountable. If you can't find a plan, that's a good reason to ask, "Why not?"

Act

Take one action that supports local wildlife. You could plant pollinator-friendly flowers, put up a birdhouse, or join a local cleanup that protects animal habitats. If you have a garden or balcony, choose native plants—they provide food and shelter for local insects and birds.

Your Stewardship Steps:
- Support local native plants in your garden.
- Avoid products that harm wildlife (e.g., unsustainable palm oil).
- Volunteer with or donate to conservation groups.

Stewardship Journal

What did I learn?

_____

_____

_____

_____

_____

_____

What small action will I try this week?

_____

_____

_____

_____

_____

_____

With whom will I share what I have learned?

_____

_____

_____

_____

_____

_____

Chapter 4
Pollution and Plastic Waste

How Waste Moves Around the World and Comes Back to Us

The Hidden Journey of Trash

When we throw something *away*, it feels like the end of the story. But, for most waste, it's just the beginning of a long, often invisible journey—one that crosses borders, pollutes ecosystems, and sometimes circles right back to us.

Your old phone, for example, might be shipped to another country where workers dismantle it by hand, often without safety protections. A plastic bottle you toss in a blue bin could be exported to a factory across the world—or worse, dumped illegally. Even your *compostable* packaging may not break down if it ends up in the wrong system.

We live in a global waste economy, where trash becomes someone else's problem. For decades, richer countries have exported plastic, electronic, and hazardous waste to poorer nations. These receiving countries often lack proper recycling or disposal infrastructure, which means waste gets burned, buried, or washed into rivers and seas.

In 2018, China—once the largest importer of global recyclables—said, "No more!" to contaminated plastics and paper. Other countries followed. Suddenly, the waste that wealthier nations used to ship overseas had nowhere to go. That exposed a harsh truth: our waste systems were designed to conceal the problem, not resolve it.

From Streets to Sea, and Back Again

A plastic wrapper dropped on a city street can travel surprisingly far. Blown into a storm drain, it may flow into a river, then out to sea. There, it joins millions of tons of plastic waste drifting in ocean currents. Some end up on remote beaches. Some breaks down into microplastics—tiny fragments that are now found in fish, shellfish, salt, and even drinking water.

Scientists have discovered microplastics in deep ocean trenches, Arctic ice, and within human bodies. We're quite literally eating and breathing the products of our own throwaway culture.

And it's not just plastics. Waste incinerators release toxins into the air. Landfills leak chemicals into groundwater. E-waste dumps in countries like Ghana and Indonesia expose children and workers to lead, mercury, and other dangerous substances. Our waste doesn't just disappear—it reshapes ecosystems and harms human health, often far from where it started.

In a globalized world, waste boomerangs. What we send away can return to us through the food we eat, the water we drink, and the air we breathe.

Unfair Burdens—Environmental Injustice

One of the most troubling aspects of the global waste system is its unfair impact on people. Poorer countries and marginalized communities often bear the worst of it—even though they typically produce the least waste.

In many places, people living near landfills or waste dumps suffer higher rates of asthma, cancer, and birth defects. Waste pickers, who often work without formal protection or recognition, sort through trash in dangerous conditions to survive.

This isn't just an environmental issue—it's a human rights issue. It's about who gets to live in a clean, safe environment and who doesn't.

But communities around the world are pushing back. Some countries are banning plastic imports. Cities are adopting zero-waste strategies. Youth movements are demanding justice, transparency, and a fundamental shift toward sustainability.

Rethinking the System

The real solution isn't just better recycling—it's producing and consuming less in the first place. That means redesigning products so they last longer. Creating systems where reuse is easier than disposal. And shifting from a *take-make-consume-throw away* model to a circular economy, where materials are used again and again.

As individuals, we can't control the entire global system—but we *can* reduce our own waste, support policies that hold corporations accountable, and choose products and companies that prioritize reuse and responsibility.

Because when we understand where our waste goes—and how it comes back—we begin to see that nothing truly goes *away*. It all stays right here, in our shared home.

Reflection: What Did You Throw Away Today?

Take a look at your trash or recycling bin right now. Do you know where that waste will go? Could anything in it have been reused, repaired, or avoided? Awareness is the first step toward change.

Action Tip: Waste Less, Matter More

Action Tip:
- Refuse what you don't need.
- Reuse what you can.
- Repair what's broken.
- Rethink what you buy.
- Recycle—but only when you know it will truly be processed.

Start with one habit this week: bring a reusable container, repair an item, or opt for plastic-free cutlery.

Case Studies in Pollution and Waste Management

Case Study: Rwanda's Bold Plastic Ban[18]

In 2008, Rwanda made international headlines by taking a bold step: it banned plastic bags completely. No plastic shopping bags, no wrapping film, no exceptions. The reason? Plastic waste was choking rivers, littering roadsides, and harming livestock and crops.

Initially, the law was difficult to enforce. However, over time, with strong public education, support for alternative packaging, and clear penalties, Rwandans gradually adopted the change. Today, Kigali—the capital—is known as one of the cleanest cities in Africa. Markets use paper bags, banana leaves, or cloth sacks instead of plastic. Monthly *Umuganda* cleanup days bring communities together to maintain clean public spaces.

The plastic ban is part of a bigger vision: a clean, green, and sustainable country. Rwanda's success shows that bold policies, combined with public engagement, can transform a nation's habits—and protect people, wildlife, and water systems in the process.

Reflection: Leadership and Willpower

What's one policy or change your city or country could make to reduce waste? Would you support it if it meant adjusting your habits?

Action Tip: Bring the Ban Home

Action Tip:
- Refuse plastic bags, wrap, and packaging whenever possible.
- Carry reusable bags, containers, and utensils to reduce waste.
- Support local businesses that offer plastic-free options.
- Share stories like Rwanda's to inspire change in your own community.

Case Study: Zero-Waste Cities—San Francisco, Kamikatsu, and Beyond[19]

In San Francisco, USA, city leaders set a goal to become zero-waste and made significant moves to achieve it. The city now diverts over 80% of its waste from landfills through composting, recycling, and policies that require businesses to sort their waste. Restaurants must use compostable packaging. Residents can compost food scraps curbside. The city even bans certain products that can't be recycled or reused.

Halfway across the world, the tiny Japanese village of Kamikatsu took an even more hands-on approach. With just 1,500 residents, the town has more than 45 different waste categories—from aluminum to bottle caps and chopsticks—and nearly everything is sorted and reused, or recycled. What they don't produce in convenience, they gain in awareness, community, and pride.

These cities prove that zero-waste isn't just a dream—it's a direction. The goal isn't perfection, but progress. And progress starts with smart design, local leadership, and people willing to do their part.

Reflection: What Could Your City Do Differently?

If you could redesign one part of your city's waste system, what would it be? Easier composting? Less packaging in stores? A reuse centre?

Action Tip: Start a Zero-Waste Habit

Action Tip:
- Pick one new habit: compost food scraps, shop in bulk, or reuse containers.
- Talk to your municipality or workplace about waste reduction programs.
- Visit a zero-waste store or refill shop if one is available near you.

Case Study: Germany's *Pfand* Deposit System[20]

Germany has one of the world's most effective recycling systems, thanks, in part, to its national *Pfand* (deposit return) program. When you buy a drink in a plastic or glass bottle or can, you pay a small deposit—usually around €0.25. When you return the container to a reverse vending machine or store, you get your deposit back.

The system is simple, convenient, and widely accepted. As a result, Germany recycles over 90% of its beverage containers. The bottles that are reused are washed and refilled up to 50 times, drastically reducing waste, pollution, and energy use.

The secret isn't just technology—it's that people are invested in it. By making packaging valuable again, Germany transformed waste into a resource and made reuse a normal part of everyday life.

Reflection: Value in a Bottle

What if every single-use item came with a refundable deposit—would you treat it differently?

Action Tip: Support Deposit Return

Action Tip:

- Support policies that bring deposit return systems to your region.
- Buy drinks in reusable or returnable containers whenever possible.
- Encourage businesses to offer refills instead of disposables.

Case Study: Plastic-Free Sari-Sari Stores in the Philippines[21]

In the Philippines, *sari-sari stores* (small neighbourhood shops) are everywhere—and so is plastic packaging. To make consumer goods affordable, companies often sell products in sachets comprising small, single-use plastic packets of soap, shampoo, coffee, and snacks.

But sachets are nearly impossible to recycle and are a major source of litter and marine plastic pollution.

Now, some communities—supported by environmental groups like Gaia South and local governments—are reimagining sari-sari stores. They're offering refill systems instead: customers bring reusable containers to refill dish soap, cooking oil, or snacks. Some stores use bulk dispensers, reusable jars, or community-led refill kiosks.

This is more than just a shift in packaging—it's a shift in mindset. Plastic isn't inevitable, and convenience doesn't have to come at the expense of the planet.

Reflection: Convenience or Care?

Are there *sachets* in your life—things you buy out of habit that could be replaced with reusable options?

Action Tip: Embrace the Refill

Action Tip:

- Find refill or bulk stores near you.
- Support businesses that allow you to bring your own container.
- Avoid mini-packs, sachets, and *sample sizes* whenever possible.

Case Study: Kenya's Plastic Bag Ban[22]

In 2017, Kenya passed one of the strictest plastic bag bans in the world. Manufacturing, selling, or using plastic shopping bags became illegal, with severe fines or imprisonment for those who violate the law. The move came after years of plastic pollution clogging drains, choking animals, and harming agriculture.

The impact was dramatic. Markets and shops quickly shifted to alternatives, such as cloth bags, woven baskets, and boxes. The amount of plastic bag litter in streets, rivers, and trees has decreased significantly. Although enforcement has had its ups and downs, public awareness has grown—and Kenya has become a global leader in taking strong action against single-use plastics.

The ban wasn't perfect, but it was powerful. It demonstrated that decisive policies, combined with public education, can alter habits and reshape norms within a few years.

Reflection: Bold Moves

Would you support a ban on plastic bags or packaging where you live—even if it meant adjusting how you shop?

Action Tip: Ditch the Bag

Action Tip:
- Carry your own reusable bag everywhere.
- Say no to plastic packaging when you have the choice.
- Support policies that ban or limit single-use plastics.

Case Study: Sweden Turns Trash into Energy—Carefully[23]

Sweden recycles or reuses nearly all of its household waste—and what's left over is often used for energy. Through high-tech incineration plants, non-recyclable waste is burned in a controlled way to generate heat and electricity, supplying millions of homes.

While waste-to-energy is a controversial topic, Sweden pairs it with strict air quality controls, robust recycling programs, and public accountability. The country's goal remains to reduce waste, not just burn it, but it views waste-to-energy as a bridge in its efforts to build a more circular economy.

Sweden even imports some waste from other countries—not because it wants to burn more trash, but because its own recycling and composting rates are so high, there's little domestic waste left to fuel its plants.

Reflection: Managing What's Left

What happens to your trash after it's collected? Could your community do more to prevent or re-purpose it?

Action Tip: Prevent Before You Burn

Action Tip:
- Focus on reducing and reusing before recycling or disposing of.
- Discover where your waste ends up—landfill, incinerator, or elsewhere.
- Support local composting or zero-waste initiatives before waste becomes a source of energy.

Case Study: Chile's *Extended Producer Responsibility* Law[24]

In 2016, Chile passed a groundbreaking law that shifted responsibility for waste away from consumers—and toward producers. Known as Extended Producer Responsibility (EPR), the law requires companies to take back, recycle, or properly dispose of the waste generated by their products, including packaging, electronics, tires, and batteries.

The idea is simple: if you make it, you're responsible for what happens when it's thrown away. This pressure has led many companies to design smarter packaging, establish recycling partnerships, and participate in national waste management networks.

The law is part of Chile's broader efforts to reduce landfill use, promote a circular economy, and hold corporations accountable for their environmental impact. It demonstrates how policy can transform an industry—and how shared responsibility is crucial to long-term sustainability.

Reflection: Who's Responsible?

When you buy a product, who do you think should be responsible for its waste—you, the store, or the company that made it?

Action Tip: Be a Smart Consumer

Action Tip:
- Look for products with minimal or recyclable packaging.
- Support companies that offer take-back or refill options.
- Encourage local businesses to carry products made with reuse in mind.

Case Study: Australia's Container Deposit Revolution[25]

Australia has embraced container deposit systems nationwide—but it wasn't always that way. Just a decade ago, only South Australia had one. Today, nearly every state and territory has a program where people earn 10 cents for returning drink containers to designated centres.

The results have been impressive. In states such as New South Wales and Queensland, recycling rates have increased significantly, and litter levels have decreased substantially. Communities benefit too—many schools and charities raise funds by collecting bottles and cans.

By placing a real value on *waste*, Australia has demonstrated that even small incentives can lead to significant environmental change.

Reflection: The Power of 10 Cents

What if a small refund could turn litter into savings—would it change how you (or your kids) treat bottles and cans?

Action Tip: Return, Reuse, Repeat

Action Tip:
- Participate in your region's deposit-return program, or advocate for one.
- Organize a community cleanup and donate the collected containers to a local charity or cause.
- Choose refillable drinks where available.

Case Study: Jordan's Waste Crisis Sparks Youth Innovation[26]

Jordan faces a growing waste management challenge—especially in its rapidly expanding urban areas. But young innovators across the country are turning that challenge into an opportunity.

In Amman, university students have launched small businesses that collect recyclables door-to-door, build eco-bricks from plastic bottles, and turn food scraps into compost. Local NGOs have supported schools in establishing zero-waste classrooms, and businesses are now offering bulk bins and plastic-free packaging options.

What's driving it all? A growing awareness that the old system—*throw it away and forget it*— is no longer sustainable. These youth-led efforts show that even in regions with limited infrastructure, people can take the lead in building a circular, low-waste future from the ground up.

Reflection: From Problem to Possibility

What local waste issue could be turned into a business, education program, or community effort where you live?

Action Tip: Support Youth-Led Change

Action Tip:
- Donate to or volunteer with youth-led environmental groups.
- Share local innovation stories on social media.
- Encourage schools and universities to teach waste literacy and creative reuse.

Case Study: Plastic-Free Islands in Indonesia[27]

Indonesia is one of the world's largest sources of ocean plastic pollution—but it's also home to some of the most inspiring efforts to stop it. Across the country's 17,000+ islands, local communities are leading the way toward a plastic-free future, starting from the ground up.

One powerful example comes from Bali, where two teenage sisters, *Melati* and Isabel *Wijsen*, launched the *Bye Bye Plastic Bags* movement in 2013. What started as a school project quickly grew into a national campaign. The sisters rallied students, organized beach cleanups, gave TED Talks, and lobbied government leaders—eventually helping to influence Bali's 2019 ban on single-use plastics.

Today, Bali is just one of many islands making change. On the remote island of Sumba, local women's cooperatives operate plastic-free markets and teach people how to make reusable bags from natural materials, such as banana fibre. In *Raja Ampat*, a marine biodiversity hotspot, dive shops, tourism operators, and village councils have banned single-use plastics to maintain clean reefs and healthy fish populations.

These changes are not always easy. Infrastructure is limited, alternatives can be expensive, and enforcement varies. However, the momentum is real—driven by local pride, youth activism, and the understanding that plastic pollution harms tourism, fishing, and public health.

Indonesia's plastic-free islands show how community leadership, cultural respect for nature, and grassroots action can create a real, visible impact—even in places that once seemed overwhelmed by waste.

Reflection: Local Action, Global Ripple

Imagine your own town or island committing to a plastic-free future. Who would need to be involved—youth, elders, shopkeepers, and schools? What role could *you* play in making it happen?

Action Tip:
- Organize or join a beach, park, or river cleanup in your area.
- Refuse plastic bags, straws, and sachets—even when they're *free*.
- Support businesses, brands, and communities working to reduce single-use plastics.
- Share inspiring local stories, such as *Bye Bye Plastic Bags*, to spread hope and inspire action.

Case Study: Palau's Zero-Tolerance for Ocean Waste[28]

Palau, a small island nation in the Pacific Ocean, is renowned for its stunning coral reefs and crystal-clear waters—but it's also gaining recognition for its ambitious conservation efforts. In 2020, Palau launched the world's first *eco-pledge* required for all visitors to sign upon arrival. Tourists promise to act in an environmentally responsible manner, including refraining from using plastic bags, straws, and harmful sunscreens that harm marine life.

Palau's government also enforces strict rules on the use of single-use plastics. Styrofoam, plastic shopping bags, and plastic utensils are banned. Schools teach ocean conservation. Locals run reef-safe shops and refill stations to reduce plastic packaging.

Why so serious? Because Palau understands that plastic pollution isn't just ugly—it threatens marine life, food security, and the tourism economy. On an island, nothing is truly *thrown away*. Every bottle, wrapper, or broken flip-flop ends up buried, burned, or floating in the sea.

Palau's approach is a reminder that small nations can lead with big ideas—and protect the places they call home by prioritizing people and the planet.

Reflection: What Would You Promise to the Planet?

If you had to write your own *Earth Pledge*, what would it include? What habits would you commit to changing to protect the land, air, or sea?

Action Tips:
- Refuse single-use plastics while travelling—pack a reusable kit.
- Choose reef-safe products and environmentally responsible destinations.
- Support tourism that gives back to nature, rather than taking from it.

The Life of a Plastic Bottle: A Waste Story

Imagine the following:

You grab a cold drink in a plastic bottle on a hot day. You finish it, then toss it in the blue recycling bin and forget about it.

But what happens next?

The bottle is collected and taken to a sorting facility. If it's clean and the right type of plastic, it might get recycled—melted down and turned into something new. However, if it's contaminated with food or mixed with the wrong materials, it may be rejected.

If that happens, it might be shipped overseas—to countries that accept plastic waste. There, it could be burned for energy (releasing toxins), dumped in open landfills, or discarded into rivers.

Eventually, the plastic breaks down into microplastics. Fish mistake it for food. So do seabirds. It enters the food chain. Months later, it may end up back in your kitchen—inside the fish on your plate.

What started as a quick sip on a summer day has become a global pollutant—one that could linger for hundreds of years.

Reflection: One Bottle's Journey

Think about how many plastic bottles you use in a week. Could you avoid them with a reusable one? What story would you rather tell?

Action Tip: Write a New Story

Action Tip:
- Carry a reusable bottle wherever you go.
- Say no to bottled drinks unless necessary.
- If you do use plastic, rinse and sort it properly to improve the chances of true recycling.

Lessons from the Front Lines of Waste Reduction

Across the world—from plastic-free beaches in Bali to deposit programs in Berlin—communities, cities, and countries are showing us what's possible. These places face different challenges, speak different languages, and live within different systems—but they share a common understanding: **we cannot continue to live in a world where everything is disposable.**

Here's what the most successful waste-reduction efforts have in common:

1. **They empower people, not just systems.**

   Whether it's schoolkids collecting bottles or communities launching refill stores, real change starts with people who care.

2. **They work with culture, not against it.**

   From banana-leaf packaging in Asia to traditional baskets in Africa, solutions that honour local knowledge are more likely to stick.

3. **They focus on prevention, not just cleanup.**

   Cleanups are powerful—but stopping waste at the source is even better. Bans, redesign, and reuse are essential.

4. **They build partnerships.**

   Cities, businesses, nonprofits, and citizens all have a role. No one can solve waste alone.

5. **They inspire through action.**

   People follow examples, not lectures. Visible, practical solutions spark ideas and hope.

The lesson is clear: we already have many of the tools we need. What we need most now is the will—and the shared belief that our everyday choices matter.

Reflection: What's Your Waste Story?

What changes have you already made in how you deal with waste? What's one more you could try this month—and who could you inspire along the way?

Forms

The Waste-Wise Pledge

I recognize that nothing truly goes *away*. What I throw out, I take responsibility for.

I pledge to:

- Reduce what I consume, choosing reusable and long-lasting items
- Refuse single-use plastics whenever I can
- Reuse, repair, and share before I replace
- Recycle properly—and only when it makes sense
- Compost my food scraps to return nutrients to the Earth
- Learn where my waste goes, and help others do the same
- Be mindful, not perfect—and keep moving forward

Because a healthier planet begins with me.

(Sign) _____

(Date) _____

Checklist - 10 Steps to Waste Less—Starting Now

1. **Refuse single-use items** (plastic bags, cutlery, straws, sachets). ☐
2. **Carry a reusable kit** (bag, bottle, container, utensils). ☐
3. **Compost food scraps**—at home or through a local program. ☐
4. **Buy less**—choose durable, repairable products. ☐
5. **Shop in bulk or refill at stores** to reduce packaging waste. ☐
6. **Repair or donate items** instead of discarding them. ☐
7. **Recycle properly**—rinse containers, follow local rules. ☐
8. **Support deposit systems** and zero-waste policies. ☐
9. **Educate yourself and others** about where waste really goes. ☐
10. **Join a cleanup or community reuse initiative.** ☐

Challenge: Check off one new habit each week for the next 10 weeks. Share your progress with a friend, family member, or online to inspire others.

Your Stewardship Steps

Reflect

Close your eyes and imagine the wildlife where you live—birds, insects, squirrels, fish in nearby rivers, native plants. Which species do you notice most often? Which ones have you seen less of lately? Think about the last time you saw a bee, butterfly, or frog. These small observations remind you that every species plays a crucial role in maintaining nature's balance—and that your daily choices can help them thrive.

Learn

Look up whether any species in your area are threatened or endangered. Is there a rare bird, insect, or plant that needs protection? You can often find this info on local government websites or nature organization pages. Learning about a local species—its habitat, the threats it faces, and the organizations working to help it—makes biodiversity feel real and close to home.

Act

Pick one simple waste-reducing habit to try this week. Swap plastic bags for reusable ones, bring your own cup for coffee, or plan meals to reduce food waste. You could also challenge yourself or your family to have one plastic-free day—it's eye-opening and fun to see how creative you can be.

Stewardship Journal

What did I learn?

_____

_____

_____

_____

_____

_____

What small action will I try this week?

_____

_____

_____

_____

_____

_____

With whom will I share what I have learned?

_____

_____

_____

_____

_____

_____

Chapter 5

Forests and Land

More Than Lungs—A Living System

Forests are often called the *lungs of the Earth*—and it's true they help regulate oxygen and carbon dioxide. But forests do much more than breathe. They are living, complex ecosystems that keep the planet in balance.

They are carbon sinks—storing massive amounts of carbon in their roots, trunks, leaves, and soil. When left intact, forests help cool the planet. When cut down or burned, they release stored carbon into the atmosphere, accelerating climate change.

But forests don't just influence the air. They shape the weather, stabilize the water cycle, prevent floods, and even affect rainfall patterns far beyond their borders. In fact, parts of the Amazon help carry moisture all the way to the United States.

Forests are home, too. Nearly 80% of Earth's terrestrial species live in forests—from tiny insects to great apes. And, for over a billion people, especially Indigenous communities, forests are also a direct source of food, medicine, culture, and identity.

Forests aren't just *natural resources*. They are living, breathing systems that support life itself—including ours.

Water Keepers of the World

Forests play a critical but often overlooked role in water security. Trees act like giant sponges—absorbing rain, filtering it through roots and soil, and slowly releasing it into streams and underground aquifers. This natural filtration system keeps water clean, cool, and flowing.

In mountainous areas, forests help stabilize slopes and reduce the likelihood of landslides. In tropical regions, tree canopies protect the soil from the beating force of rain, preventing erosion and keeping nutrients in place.

Forests even influence rainfall. Through a process called transpiration, trees release water vapour into the air—helping to form clouds and encourage rain. When forests disappear, so does this cycle. That's one reason why deforestation can lead to droughts in nearby farmlands and cities.

In places like the Congo Basin or the Amazon, vast *rainforests* truly earn their name—not just for the rain they receive, but for the water they help create and move.

If we want secure, reliable water sources for farms, cities, and future generations, protecting forests is one of the smartest things we can do.

Forests as Homes and Healers

Forests are among the richest habitats on Earth. Tropical rainforests, boreal forests, temperate woodlands—each hosts a unique cast of species adapted to its rhythms. Birds, mammals, fungi, amphibians, pollinators—all are part of an intricate, ancient balance.

But forests are also medicine cabinets. Many of the pharmaceuticals we use today—from aspirin to cancer treatments—come from forest plants. Scientists estimate we've studied only a fraction of tree and plant species for their medicinal potential. When we lose forests, we lose possible cures.

For Indigenous peoples, forests are not only life-sustaining—they are sacred. They hold stories, identities, spiritual practices, and traditional knowledge that have been passed down through generations. Many Indigenous communities have managed forests sustainably for centuries, with wisdom about when to harvest, when to leave an area fallow, and how to coexist with wildlife.

Conserving forests means honouring this knowledge—and recognizing that human health, cultural heritage, and biodiversity are deeply intertwined.

When Forests Fall—The Cost of Deforestation

We lose roughly 10 million hectares of forest each year—an area about the size of Portugal. Much of this is driven by agriculture (especially beef, soy, and palm oil), mining, road building, and logging—both legal and illegal.

Deforestation doesn't just remove trees—it unravels whole ecosystems. Wildlife loses its home. Soil loses its structure. Carbon floods the atmosphere. And communities lose clean water, food sources, and protection from storms and heatwaves.

Some forests can regrow, but recovery takes decades—and many ecosystems never fully return. Meanwhile, *tree planting* alone is not a substitute for old-growth forests. Reforestation is important, but protecting what we still have is even more urgent.

The Forest Future—Protect, Restore, Respect

The good news? People around the world are fighting to protect forests—and win.

Indigenous communities in Brazil are defending the Amazon. Schoolchildren in Uganda are planting trees along riverbanks. Governments are signing global forest pledges, and businesses are starting to reassess their supply chains to prevent deforestation.

We all have a role to play:

- Support forest-friendly products. Look for FSC(Forest Stewardship Council)-certified paper and wood, or palm oil from sustainable sources.
- Cut back on meat from deforestation-linked regions.
- Donate to or volunteer with reforestation and land protection groups.
- Plant native trees, and protect those already growing in your neighbourhood.

Forests are more than lungs. They are Earth's beating heart—regulating the climate, nourishing life, and reminding us of our deep connection to the natural world. To care for forests is to care for our collective future.

Case Studies: Forest Stewardship in Action

Case Study: Community Forests in Nepal[29]

Nepal has become a global leader in community-based forest management. Since the 1990s, over 22,000 *community forest user groups*—comprising local villagers—have assumed management of degraded forests.

These groups decide how to harvest firewood, protect wildlife, prevent illegal logging, and share benefits. They plant native species, build firebreaks, and monitor forest health. In many areas, forest cover has increased, and biodiversity has rebounded—all while generating income, creating jobs, and fostering stronger local leadership.

Nepal's model shows that when people are empowered to manage the land they depend on, forests can thrive.

Case Study: Indigenous Guardians of the Amazon[30]

In Brazil, Indigenous territories comprise less than 14% of the Amazon rainforest, yet they account for over one-third of its intact forests. Studies show that deforestation rates inside Indigenous lands are dramatically lower than in surrounding areas.

Groups like the *Kayapo*, *Yanomami*, and *Munduruku* patrol their territories, monitor illegal logging, and use satellite tools to report environmental crimes. Their stewardship protects biodiversity, carbon stores, and water systems—not just for Brazil, but for the planet.

Protecting Indigenous rights is not only a matter of justice—it's one of the most effective climate solutions we have.

Forms

The Forest Keeper's Pledge

I pledge to stand with forests—as a protector, a learner, and a mindful consumer.

I will reduce my use of products that lead to deforestation.

I will support the people who live with and care for forests.

I will speak up for trees that have no voice.

I will value forests not just for what they give—but for what they are.

I am part of the forest's story—and I choose to help it grow.

(Sign) _____

(Date) _____

Your Stewardship Steps

Reflect

What role do forests play in your life—even if you don't live near one? Think about the paper you use, the food you eat, and the climate you rely on. What could you change to tread more lightly?

Learn

Understand how forests work as ecosystems—not just tree cover, but soil, fungi, wildlife, and water working together. Research your region's native trees, the threats to its forests, and local conservation efforts.

Act

- Support reforestation projects.
- Eat less meat (ties to land use).
- Plant or protect native trees in your area.
- Choose FSC-certified paper and wood products.
- Choose sustainably sourced wood and furniture.
- Avoid *fast furniture* and short-lived wood products.
- Support Indigenous forest rights and protection campaigns.
- Every tree you help protect or plant is a gift to the planet's lungs.
- Consider donating to a reputable reforestation project or volunteering for a local tree-planting event.
- Reduce or eliminate products linked to deforestation (e.g., palm oil, soy-fed meat).

Stewardship Journal

What did I learn?

_____

_____

_____

_____

_____

_____

What small action will I try this week?

_____

_____

_____

_____

_____

_____

With whom will I share what I have learned?

_____

_____

_____

_____

_____

Part III

Making Stewardship a Way of Life

Chapter 6

Everyday Actions at Home

Energy, Water, Waste, Food: The Practical Low-Hanging Fruit
Energy—Use Less, Use Smarter

Energy powers everything in our homes—lights, appliances, phones, heating, and cooling. But most electricity still comes from fossil fuels, so using less saves money *and* reduces emissions.

The low-hanging fruit:

- **Switch to LEDs.** They use up to 90% less energy than old bulbs—and last longer too.
- **Unplug devices when not in use.** Many electronic devices draw power even when turned *off*.
- **Use a power bar.** Flip it off to cut vampire energy from chargers and appliances.
- **Set thermostats wisely.** Lower in winter, higher in summer—even a 1°C change saves energy.
- **Hang dry clothes.** Clothes dryers are among the biggest household energy hogs.
- **Upgrade insulation and windows.** A one-time effort that pays off for decades.

If your utility offers renewable energy options, consider signing up for them. If you rent, discuss energy upgrades with your landlord—or focus on what you can control, such as using efficient bulbs and practicing smart power use.

Water—Use Every Drop Wisely

Fresh, clean water is more precious than most people realize—and climate change is making droughts more common. Fortunately, there are simple ways to conserve water at home without compromising comfort.

The low-hanging fruit:

- **Fix leaks.** A dripping tap can waste hundreds of litres per year.
- **Install low-flow shower heads and aerators.** Big savings, small cost.
- **Turn off the tap while brushing your teeth.**
- **Use full loads only.** Run dishwashers and laundry machines only when full.
- **Collect rainwater for gardens.** Free water, no hoses needed.
- **Use mulch and native plants outdoors.** They need less watering and are more resilient.

These small shifts add up—and they remind us that water, like energy, is a shared resource, not an unlimited one.

Waste—Shrink Your Footprint, Not Your Lifestyle

Every piece of waste has a backstory: energy to make it, water to process it, carbon to transport it—and a future that often leads to a landfill, incinerator, or the ocean.

The goal isn't perfection—it's progress. Start where it's easiest for you.

The low-hanging fruit:

- **Refuse single-use items**, such as plastic bags, straws, and disposable cutlery.
- **Switch to reusable** items—such as cloth napkins, metal bottles, and glass containers.
- **Compost food scraps.** It reduces landfill methane and creates rich soil.
- **Sort recycling properly.** A little care makes a big difference.
- **Buy less.** The less you bring in, the less you throw out.
- **Repair instead of replacing.** A screwdriver or a wrench can keep stuff in use.

Waste reduction isn't about deprivation—it's about thoughtful living and knowing that every item we keep out of the bin is a win for the planet.

Food—Eat for the Planet (and Your Health)

Food is one of the most powerful levers for climate action. It impacts land use, water resources, emissions, and biodiversity. And the good news is: we eat multiple times a day, which means we have daily opportunities to make a difference.

The low-hanging fruit:

- **Waste less food.** Plan meals, store food properly, and love your leftovers.
- **Eat more plants.** Even one meatless day a week helps reduce land and water use.
- **Buy local and seasonal.** It cuts transport emissions and supports small farms.
- **Choose sustainable seafood** and certified products when possible.
- **Grow something.** Even herbs on a windowsill reconnect us with the source of our food.

You don't have to change everything overnight. Just start with what feels doable—and build from there.

Putting It All Together—Small Steps, Big Impact

The best part of low-hanging fruit is that it's… low. Easy to reach. Fast to act on. It doesn't require solar panels or a zero-waste pantry—just a little attention and intention.

Try this:

- Pick *one small action* each week from each category—energy, water, waste, food.
- Keep track in a journal or checklist.
- Celebrate progress—even tiny wins matter.

None of us can solve the environmental crisis alone—but each of us can be part of the solution. By tending to our own homes with care, we create ripples of change that stretch beyond our walls.

Case Study: The Low-Impact Lewis Family (Canada)

The Lewis family lives in a small home outside Ottawa. They're not activists, off-grid, or perfect—just a family of four trying to live more lightly on the Earth. Here's what their practical, low-impact life looks like:

- **Energy:** They swapped all bulbs for LEDs, unplugged electronics when not in use, and dry laundry outdoors when the weather allows. In winter, they use thick curtains and wear sweaters before turning up the heat.
- **Water** conservation measures, such as low-flow shower heads, a rain barrel for the garden, and mulching around native plants, reduce water use year-round.
- **Waste:** They compost all food scraps, use reusable shopping bags and containers, and buy in bulk with minimal packaging. They've even learned basic sewing to repair clothing.
- **Food:** The Lewis family primarily eats vegetarian meals, grows herbs and tomatoes on their balcony, and plans their weekly meals to minimize waste. They buy from a local farm in the summer.

They still use a car, order pizza on tired nights, and occasionally use plastic wrap—but they focus on doing their best, not being perfect. Over the past five years, they've cut their energy use by 20%, reduced waste by two-thirds, and inspired their neighbours to join in.

Their motto? *"If it's good for the planet and doable for us, we'll give it a try."*

Your Stewardship Steps

Reflect

What small habits are costing the Earth more than you realized? What would your home look like if it aligned better with your values?

Learn

Understand your household's impact—where your energy comes from, how much food you waste, and what happens to your trash and water.

Act

Pick one small change each week using the Home Action Tracker.

Talk with family or roommates—shared goals are easier to reach.

Support local solutions (farmers' markets, repair cafés, water-saving rebates).

Forms

Home Action Tracker: Small Steps, Big Impact

	Energy	Water	Waste	Food
Action 1				
Action 2				
Action 3				
Action 4				

Instructions: *Choose one small action from each category each week.*

Fill in the box when you have completed it. Track your progress for a more sustainable home!

The Home Steward's Pledge

I pledge to care for my home—not just the building, but the planet it depends on.

I will reduce waste, save energy and water, and choose food that nourishes both people and planet.

I will start with what I can—and grow from there.

Because every home, including mine, is part of the Earth's future.

(Sign) _____

(Date) _____

Family Habits and Kids as Stewards Too

Why Families Matter

Our homes are the first place we learn how to treat the world. That includes not just each other but the Earth itself. The habits we build as families—around food, energy, water, and waste—send a powerful message about what we value and why it matters.

And while many environmental problems may seem large and distant, the truth is that real change often begins at the dinner table, in the laundry room, or on a walk to school. When families shift their habits together, kids grow up with a natural understanding of stewardship—not as a chore, but as a way of life.

The good news? Kids are naturals. They care deeply about animals, fairness, and the future. When given the tools and encouragement, children can become powerful voices for change—and even teach the adults around them a thing or two.

Habits That Stick—Make It Easy, Make It Fun

Creating a more eco-friendly household doesn't have to mean strict rules or big sacrifices. The key is to make new habits simple, visible, and, especially with kids, a little fun.

Ideas for energy-conscious families:

- Let kids be *light monitors* who turn off switches when leaving rooms.
- Use stickers or colourful tape on power strips to show which ones should be turned off.
- Make a game out of reducing screen time and doing more activities that use no power at all.

Ideas for water-wise families:

- Time showers together with a song; when it ends, water off!
- Use watering cans instead of hoses to teach care and conservation.
- Let kids help collect rainwater or reuse leftover drinking water for plants.

Ideas for low-waste living:

- Set up easy-to-use recycling and compost bins with clear labels or colour codes.
- Let kids decorate reusable containers, bags, or lunch kits.
- Conduct a weekly *waste audit* to determine what gets thrown out and how we can improve.

Food-smart tips for families:

- Cook meals together and talk about where food comes from.
- Grow herbs or veggies in pots, even on a windowsill.
- Let kids help plan meals or choose a *no-waste* leftover night.

When kids have ownership over small actions, they learn agency and that their choices have real power.

Kids as Earth Heroes

Children love stories, and the story of Earth needs young heroes.

Some of the world's most inspiring environmental leaders began their journeys as children. Think of Greta Thunberg's school strike, or the youth-led *Bye Bye Plastic Bags* movement in Bali. But even smaller, quieter actions matter; a child who teaches their class how to compost or helps plant trees in the neighbourhood is writing a powerful story of their own.

Ideas to empower kids:

- Create an *eco-jobs* chart at home (compost helper, laundry line assistant, garden waterer).
- Watch nature documentaries together and talk about solutions.
- Visit parks, forests, or farms to establish genuine connections with nature.
- Help kids write letters to local leaders or businesses about environmental concerns.
- Start a school or neighbourhood cleanup, even with just a few friends.

The goal isn't to pressure kids to *save the planet*; it's to help them feel connected, capable, and part of something meaningful.

Talk About It. The Power of Family Conversations

Sometimes, the most important thing families can do is simply talk to each other. Ask questions, share hopes, express frustration, and imagine alternatives. These conversations shape how kids think about their place in the world, and how they learn that caring for it is a shared responsibility.

Try these questions at dinner, in the car, or while doing chores:

- What do you love most about nature?
- What would our home look like if it were 100% eco-friendly?
- What's something we use that we could replace with a better option?
- What kind of future do you want, and what can we do now to help build it?

No one needs to have all the answers. What matters is creating space to think, wonder, and grow together as a household and as stewards of the Earth.

Forms

Family Stewardship Pledge

We pledge to care for our home, our neighbourhood, and our planet—together.
We will:

☐   Use only what we need—and waste less.

☐   Help each other save energy, water, and food.

☐   Take care of animals, plants, and wild places.

☐   Speak up for the Earth when it needs a voice.

☐   Try our best every day—and have fun doing it!

Family name: _____   Date: _____

Kids' Action Checklist

☐ Turn off the lights when I leave the room.

☐ Use a reusable water bottle or lunch container.

☐ Help sort recycling and take out compost.

☐ Water plants and help care for the garden.

☐ Pick up litter during a walk or park visit.

☐ Help plan or cook a no-waste meal.

☐ Talk to friends about helping the earth.

☐ Create art or stories about nature.

☐ Use both sides of paper before recycling.

☐ Learn one new thing about animals or plants.

Weekly Eco-Jobs Chart for Kids & Families

☐ Write your family's names in the columns.

☐ Assign eco-jobs for each day of the week.

☐ Rotate tasks daily or weekly so everyone gets a turn!

Eco-jobs	Mon	Tues	Wed	Thurs	Fri	Sat	Sun
Light Switch Monitor							
Compost Collector							
Recycling Sorter							
Plant Waterer							
Food Saver Hero							

Mini Case: A Family That Went Zero Waste

Starting with One Jar

In 2016, the Tan family, parents Nina and Ken, and their two children, stumbled upon a video of someone who could fit all their trash for a year into a single glass jar. It seemed impossible. But something about it stuck.

At the time, the Tans lived a typical suburban life outside Vancouver. They recycled, turned off lights, and brought reusable bags *sometimes*. But like many families, they were throwing out full garbage bags every week—diapers, food packaging, snack wrappers, broken toys.

"We thought we were being eco-friendly," Nina says. "But when we actually looked at our trash, we saw how much was just part of our routine. It wasn't intentional; it was invisible."

So, the Tans set a small goal: to fit one week's trash into a single container. No pressure. Just a challenge.

That first week, they failed—but they learned from their mistakes.

Progress Over Perfection

Over time, the Tans rethought one small category at a time.

Groceries? They brought mesh produce bags, shopped at a bulk store, and switched to buying in larger containers. They stopped buying individually wrapped snacks.

Toiletries? They swapped shampoo bottles for bars, tried DIY toothpaste, and bought toilet paper wrapped in paper, not plastic.

Trash? They implemented better recycling practices and started composting. They made a family game out of finding package-free alternatives. Birthdays became experiences instead of plastic-filled loot bags.

They didn't do it all at once. They had setbacks. Ken once joked that "zero waste with kids is more like 'low waste with grace.'"

However, within a year, their household garbage had shrunk to a single small bag every few months. Most of what remained was medical waste or items for which there were no local recycling options.

More importantly, the kids became stewards too—asking questions, helping sort waste, and reminding their parents to bring containers to the bakery.

The Invisible Shift

What surprised the Tans most was how their values shifted in tandem with their habits.

"We started buying less overall," Nina explains. "When you stop bringing home trash, you stop bringing home clutter too."

They began fixing broken items instead of replacing them. They chose slower weekends, spent in nature or cooking, over shopping trips. They noticed their grocery bill dropped. Their kids grew up knowing that waste isn't just something that disappears—it's something we choose to create, or not.

Going zero waste also sparked conversations with friends, neighbours, and school staff. "People didn't always understand," Ken says. "But many were curious. Some made changes just from seeing what we were doing."

They didn't evangelize. They just lived differently—and let the results speak for themselves.

What They Learned (and What They'd Say to You)

The Tans don't claim to be perfect. They still occasionally buy packaged treats. They order online sometimes. Life happens. However, their journey demonstrates that significant change can emerge from curiosity, intention, and persistence.

Their biggest lessons?

- **Start small.** Don't try to do it all at once. Pick one thing, like saying no to plastic straws, and build from there.
- **Systems matter.** Make your zero-waste setup easy to use and hard to forget.
- **Kids can lead.** Involve children early and let them be part of the process.
- **Perfection isn't the goal.** Every small reduction in trash is a win.
- **Share your story.** You never know whom you'll inspire.

"We're not saving the world on our own," Nina says. "But we're doing our part and teaching our kids that they can too."

Mini Case: The Apartment That Went Green

A Tiny Space, A Big Impact

Jasmine lives in a one-bedroom apartment in downtown Toronto. She doesn't have a backyard, a compost bin, or a car and, for years, she assumed that meant she couldn't make much of a difference.

"I thought environmentalism was for people with gardens and money," she says. "But then I realized that I still make choices every day. And those choices add up."

Jasmine began by looking around her 600-square-foot home with new eyes. Her light bulbs? Incandescent. Her trash? Overflowing with food waste and plastic takeout containers. Her electricity? From the city grid—mostly fossil fuels.

So she picked one category—energy—and got to work.

Going Greener, One Choice at a Time

Jasmine replaced every bulb with LEDs and unplugged her TV and kitchen appliances when not in use. She signed up for her electricity provider's green energy option. It cost a few extra dollars per month—but gave her peace of mind.

Then she tackled food waste. She couldn't compost in her building, but she found a local farmers' market that accepted food scraps. She kept a small container in her freezer and dropped it off each Saturday.

She also began meal planning—cooking big batches, freezing leftovers, and shopping with a purpose. Her food waste dropped significantly, as did her grocery bill.

Next came takeout: Jasmine switched to restaurants that let her bring her own containers, and joined a community tiffin program that delivered meals in reusable tins.

And on weekends? She started growing herbs on her windowsill and joined a rooftop garden cooperative across the street.

Living by Example, Not Perfection

Jasmine isn't zero-waste. She still uses the elevator. She still gets the occasional packaged snack. But she's proud of what she's changed and how far she's come.

Today, her neighbours call her the *eco wizard*. They've asked her for tips, swapped ideas, and even started their own shared compost drop-off box in the building lobby.

"I didn't preach," she says. "I just did my thing. But it's amazing how contagious small actions can be."

Lessons from One Small Apartment

Jasmine's story is a reminder that you don't need to live on a farm or own a house to live sustainably. Some of her key insights:

- **Start with what you control.** Lighting, food, and waste habits are within reach, even in a rental unit.
- **Look for local hacks.** Libraries, co-ops, community gardens, and farmers' markets often support low-waste living.
- **Use what you have.** Reusing old containers or jars can be just as powerful as buying fancy *eco-products*.
- **Share your journey.** One person's quiet shift can have a ripple effect.

As Jasmine puts it, "My apartment may be small, but my impact doesn't have to be."

Mini Case: The Three-Generation Green Home

Many Voices, One Roof

The Patel household in Mississauga is lively and full of life. Grandparents, *Meena* and *Raj*, daughter *Priya* and her husband Alex, and their two kids all share a busy two-story home.

Six people, three generations, and one shared goal: live more lightly on the Earth.

"It wasn't always a shared goal," *Priya* laughs. "My mom didn't get why I wanted to compost, and my kids didn't understand why we couldn't always buy new toys."

But little by little, the family began to find common ground, often in surprising ways.

Old Wisdom, New Ideas

Raj had grown up in a village in India where nothing was wasted. "We reused everything," he says. "Clothes, food scraps, jars, not because it was trendy, but because it made sense."

So when the family started composting and storing leftovers in glass containers, *Raj* felt at home. *Meena*, too, revived her habit of making yogurt and spice blends from scratch, thereby reducing waste and plastic use.

The kids took the lead on recycling and made colourful signs for the bins. They also started a "swap shelf" where toys, books, and clothes were exchanged instead of tossed.

Alex, an electrician, conducted an energy audit of the house, helping everyone understand where electricity was wasted and how small changes, such as weatherstripping and installing LED lights, could make a significant difference.

Every generation brought a piece of the puzzle. Together, they built a more sustainable home and a deeper respect for each other's contributions.

From Chores to Shared Responsibility

One of the family's biggest breakthroughs came from redefining chores as "Earth jobs." Composting wasn't just a smelly task; it was saving food from the landfill. Turning off the thermostat wasn't just frugality; it was climate care.

The kids got badges for *Energy Ninja* and *Water Saver*. *Meena* taught them how to sew torn clothes. Priya took everyone to a repair café to fix a broken toaster. Alex led a DIY project to install a rain barrel.

Even movie nights changed; they watched nature documentaries and held discussions over homemade popcorn.

"We didn't try to do everything," says *Priya*. "But we made it part of how we lived together, and it brought us closer."

Lessons from a Multi-generational Green Home

The *Patels*' story demonstrates how environmental stewardship can become a family culture, rather than just an individual effort.

Their takeaways:

- **Honor different strengths.** Elders may bring experience in low-waste living, while kids bring curiosity and creativity.
- **Involve everyone.** A sustainable home works best when everyone has a role, not just one *person* dedicated to sustainability.
- **Celebrate progress.** The *Patels* hang a *Planet Points* chart on the fridge and reward themselves with a family picnic after big wins.
- **Make room for learning.** Mistakes happen, but each one is a step forward.

"Our home isn't perfect," *Raj* says. "But we're learning every day and that's what matters."

Forms

Start Your Journey: Family Sustainability Planner

Draw inspiration from families around the world and use this page to design your own eco-journey. Pick one habit in each category to try this week, and build your own green home routine!

Energy
What will we try?

Water
What will we try?

Waste
What will we try?

Food
What will we try?

Community
What will we try?

Family Reflection
Why do we care about the planet? What do we hope to change together?

Our family name: _____   Date: _____

Our Family's Green Living Pledge

We, the undersigned members of the _____ family, pledge to:

- Use energy wisely and turn things off when not in use.
- Conserve water and treat every drop with care.
- Reduce waste by reusing, recycling, and composting.
- Choose foods that are kind to the earth.
- Take care of nature and speak up for our planet.
- We will learn together, grow together, and do our best, not to be perfect, but to be part of the solution. Every small action we take at home helps build a better world for everyone.

Signed by our family:

(Name)	(Name)
(Name)	(Name)
(Name)	(Name)
(Name)	(Name)

Date: _____

Our Eco-Pledge: For Classrooms, Scouts & Clubs

We, the members of our learning group, pledge to:

- Save energy by switching off lights and devices when not in use.
- Use water wisely and avoid waste.
- Reduce trash by recycling, reusing, and composting what we can.
- Choose snacks and meals that are kind to the planet.
- Take care of our schools, parks, and local nature with pride.

We will learn together, take action together, and help others join in the effort.

We know that small actions lead to big changes, and we're ready to do our part!

Group name: _____   Date: _____

Signed by group members:

_____	_____
(Name)	(Name)
_____	_____
(Name)	(Name)
_____	_____
(Name)	(Name)
_____	_____
(Name)	(Name)
_____	_____
(Name)	(Name)

Date: _____

Your Stewardship Steps

Reflect

Think about the last time you spent time in a forest, park, or green space. How did it make you feel: calmer, more grounded, connected to something bigger? Now reflect on what that patch of nature does for you, even when you're not there: it cleans the air, stores carbon, and provides a home for wildlife. This simple moment of gratitude can inspire you to protect forests and open spaces near and far.

Learn

Research whether the paper, wood, or food you use comes from sustainable sources. Check for certifications, such as FSC (Forest Stewardship Council) or Rainforest Alliance, on the products you buy. If you have a favourite product, like coffee, chocolate, or furniture, learn how it's grown, harvested, or manufactured. Small shifts in how you shop can support healthy forests and fair livelihoods around the world.

Act

Pick a *quick win* to save energy, water, or reduce waste right where you live. Turn off the lights when you leave a room. Switch to LED bulbs. Wash laundry in cold water. Use a draft stopper in winter. These tiny home habits not only reduce your footprint; they also save money.

Stewardship Journal

What did I learn?

_____

_____

_____

_____

_____

_____

What small action will I try this week?

_____

_____

_____

_____

_____

_____

With whom will I share what I have learned?

_____

_____

_____

_____

_____

_____

Chapter 7

Working With Your Community

How local action multiplies your impact

Why Local Action Matters

Big environmental problems, such as climate change, deforestation, and pollution, can feel impossibly distant. What can one person do about a melting glacier or a burning rainforest?

The answer: *a lot*, especially when acting locally.

Your neighbourhood, town, or city is where policy meets people. It's where your voice is heard most quickly, your habits are visible to others, and your energy can spread outward. What you do locally, from planting a pollinator garden to joining a town hall meeting, may seem small. But it sets off a chain reaction.

Local action:

- Makes environmental change more visible.
- Builds community and trust.
- Sparks innovation that others can copy.
- Puts pressure on higher levels of government to follow suit.

By acting locally, you're not only solving problems close to home, you're also helping shape a culture of care and resilience.

Real Examples of Local Multiplier Effects

Local action isn't theoretical; it's working everywhere. Here are just a few real-world ripple effects:

- **In the Philippines**, one coastal village banned single-use plastics after a school cleanup campaign. That decision inspired neighbouring villages, eventually pushing a regional government to introduce wider bans.
- **In California**, a community that started a free home composting program diverted thousands of tons of waste from landfills, enough to spark policy changes at the city level.
- **In Kenya**, women-led tree planting groups began as small gatherings. Today, groups like the Green Belt Movement have reforested entire landscapes and shaped global dialogue on land restoration.
- **In your neighbourhood**, a resident starts a tool-sharing shed, which reduces waste, cuts consumption, and brings neighbours closer together.

When people see others acting locally, it breaks the myth of helplessness. It shows that change is not just possible; it's happening right here and now.

Local Action You Can Take

Local action doesn't need to be loud or complicated. You can amplify your impact by starting where you are.

Connect:

- Join or create a local green team, garden group, or zero-waste challenge.
- Attend town halls, school board meetings, or city planning sessions.
- Find neighbours who share your values and brainstorm shared projects.

Contribute:

- Host a seed swap, native plant day, or litter cleanup.
- Offer to speak at a school or library about sustainability.
- Support or start a Buy Nothing group, tool library, or skill-share circle.

Influence:

- Advocate for policies such as improved transit, composting, tree protection, and green space.
- Write a letter to your local newspaper or councillor.
- Vote, and help others understand the local environmental stakes.

Even small steps, like biking to the farmers market or starting a sidewalk garden, can nudge others to reconsider their habits. You become part of a feedback loop that grows bigger with every act of courage, connection, and care.

From Local to Global, The Multiplier Effect

When you act locally, you also contribute globally in powerful ways:

- You **reduce your footprint** directly: less waste, fewer emissions.
- You **model change** others can replicate.
- You **feed networks** of inspiration that cross borders.

Think of your local action as a spark. One spark ignites another and, eventually, whole systems catch on. That's how social tipping points are reached: not by one big shift, but by many local ones joining forces.

A sustainable world will be built city by city, neighbourhood by neighbourhood, street by street, and it starts with you.

Case Studies: Community Gardens, Co-ops, Local Policy Wins

Greening Together, Community Gardens That Grow More Than Food

Case: Thorncliffe Park Urban Garden, Toronto, Canada[31]

In a densely populated neighbourhood of high-rise apartments, a group of newcomer women came together to create an unlikely oasis: a multicultural community garden. Built on a school property with city support, the garden transformed a patch of underused land into 60 thriving plots where families grow a variety of vegetables, including eggplants, herbs, and tomatoes.

But it's about more than food. The garden fosters trust across cultures, provides mental health benefits, and alleviates food insecurity. Local youth learn to grow vegetables. Elders share stories and techniques from their home countries. During harvest season, potlucks bring together neighbours who have never met before.

This garden has since inspired similar spaces in nearby neighbourhoods. The lesson? When people plant seeds together, they also grow a community.

People Power—Community Co-ops for a Greener Economy

Case: Worker-Owned Green Cooperatives, Mondragón, Spain[32]

In the Basque region of Spain, the town of Mondragón is home to one of the world's most successful cooperative economies. While not focused solely on sustainability, many of its co-ops prioritize green energy, circular materials, and local resilience.

One such co-op, *Eroski*, is a supermarket chain run by its workers and committed to sourcing locally and reducing packaging waste. Others focus on sustainable home construction, appliance repair, or energy efficiency services.

This model illustrates what can happen when communities reclaim control over their economies: profits are reinvested locally, decisions are made democratically, and sustainability becomes an integral part of the mission, not merely a marketing tool.

Today, new co-ops around the world are drawing inspiration from this model, using it to create local jobs and environmentally friendly businesses.

Small Voice, Big Change, Local Policy Wins From Ordinary People

Case: Youth Climate Lobby Pushes Green Infrastructure, Portland, Oregon[33]

In Portland, a group of high school students formed a youth climate council to push for real climate action. They partnered with local nonprofits, gathered public input, and organized town halls. Their focus? Encouraging the city to reinvest in tree planting, bike lanes, and the electrification of public buildings.

Their persistence paid off. In 2020, the city passed a new ordinance requiring climate considerations in all urban planning decisions, including building codes, road redesign, and budgeting.

More importantly, the youth were given a formal advisory role to the city council, proving that young voices, when organized and consistent, can reshape the system.

Other local wins around the world:

- A single mother in South Africa led her township to ban illegal dumping.
- A faith-based group in Australia helped pass plastic-bag bans across multiple states.
- Residents in a small German village crowdfunded and built their own solar microgrid.

What These Stories Teach Us

These case studies aren't just feel-good stories; they're blueprints for success. They remind us that:

- **Communities know their needs best.** Locals understand what will actually work in their own area.
- **Small-scale efforts often scale up.** Gardens turn into networks. Policies spread to neighbouring cities.
- **Leadership can come from anywhere.** Teachers, teens, elders, and new immigrants can all spark change.
- **Collaboration is key.** Most success stories involve partnerships between citizens, governments, and organizations.

You don't have to start a movement. Just join and/or support one that already exists. Whether it's a shared garden, a new co-op, or a local bylaw, your actions help write the next case study for someone else to follow.

Forms

Eco-Challenge Tracker

Use this page to track who participated and what actions they completed during your eco-challenge.

Name	Actions Completed	Total Points

Certificate of Participation

This certifies that

_____

has actively participated in the

Neighbourhood Eco-Challenge
In recognition of their dedication to sustainability,
community engagement, and making a positive impact on the planet.

Challenge Leader _____   Date _____

Together, we grow a greener tomorrow!

Local Action Starter Map

Use this map to discover opportunities for environmental action in your community.

Fill in ideas, organizations, or people under each category, then pick one to start with!

☐ **Local Nature Spaces**

☐ **Schools, Libraries, and Youth Groups**

☐ **Neighbours and Community Members**

☐ **Local Businesses and Organizations**

☐ **City Council or Municipal Services**

☐ **Projects I'd like to Try**

☐ **My First Local Step**

☐ When will I do it? _____

Neighbourhood Eco-Challenge Template

Start a fun and friendly challenge with your neighbourhood or local group!

Set a timeline, track progress, and celebrate eco-wins together.

Challenge Name: _____

Challenge Date: from _____ to _____

What we hope to achieve:

Eco-Actions (check all that apply):

- ☐ Reduce household waste
- ☐ Start composting
- ☐ Use less energy
- ☐ Bike and walk more
- ☐ Plant native species
- ☐ Clean up a park or street
- ☐ Share or repair instead of buying new
- ☐ Host a green event or workshop

Our Eco-Team Name (optional): _____

How we'll celebrate our progress:

Signed by (challenge leader or group):

Community Action Cards

☐ **India: Women's Waste Co-op**

In Pune, India, thousands of women formed a cooperative to collect, sort, and recycle waste. Known as the SWaCH co-op, these women provide essential waste services while diverting tons of material from landfills. They also promote composting, upcycling, and cleaner streets.

☐ **What this inspires me to try in my community:**

☐ **Brazil: *Favela* Green Roofs**

In Rio de Janeiro, residents of low-income neighbourhoods started installing green roofs to reduce indoor heat and manage rainwater runoff. These grassroots projects not only cool homes but also provide small spaces for food and native plants.

☐ **What this inspires me to try in my community:**

□ **Netherlands: Bike City Planning**

Dutch cities like Utrecht and Groningen have transformed their streetscapes to prioritize bikes over cars. With wide lanes, protected intersections, and bike parking structures, they've reduced emissions and created healthier communities, demonstrating how urban policy can respond to citizen advocacy for safer, cleaner transportation.

□ **What this inspires me to try in my community:**

□ **Uganda: Solar Sister Empowerment**

Through the Solar Sister program, women in rural Uganda are trained to distribute clean solar lights and cookstoves, empowering them to improve their lives. The initiative addresses energy poverty, conserves forests, and generates employment, demonstrating the potential of community entrepreneurship for achieving sustainability.

□ **What this inspires me to try in my community:**

☐ **Blank Template**

☐ **Case Study Title / Project Name:**

```

```

☐ **What happened? Who led it? What was the result?**

```

```

☐ **What this inspires me to try in my community:**

```

```

Your Stewardship Steps

Reflect

Walk through your home, room by room, and ask yourself: Where does the energy come from that powers your lights, heat, or appliances? Where does your water come from, and where does it go? Do you see places you're wasting energy or water without realizing it? This reflection helps you identify easy ways to reduce waste, lower your bills, and minimize your environmental footprint.

Learn

Find out what local programs or rebates exist to help you save energy or water. Some cities offer free home energy audits, discounts on energy-efficient appliances, or incentives for fixing leaks and installing more effective insulation. Even if you rent, you might discover simple ways to cut waste and lower your bills. A quick search on your city's website is a good starting point.

Act

Take a simple step to connect with others who care. Join a local cleanup, visit a community garden, or attend an event hosted by an environmental group. If nothing like that exists, start small: invite friends or neighbours to plant flowers, pick up litter, or share ideas for local improvements.

Stewardship Journal

What did I learn?

_____

_____

_____

_____

_____

What small action will I try this week?

_____

_____

_____

_____

_____

With whom will I share what I have learned?

_____

_____

_____

_____

_____

Chapter 8
Policy and Advocacy Made Simple

Why personal change and systemic change go together

False Choices: Why It's Not One or the Other

A common myth divides the environmental movement into two camps: those who focus on *individual lifestyle choices,* such as using less plastic or reducing meat consumption, and those who focus on *systemic change*, including legislation, corporate accountability, and fossil fuel divestment.

But, in truth, these aren't opposing ideas; they are **mutually reinforcing**.

- When people **change their habits**, they build awareness, set an example, and send signals to markets and policymakers.
- When **systems change**, they make it easier for individuals to live sustainably by default.

You can't have one without the other. Personal change is a powerful gateway to advocacy. Systemic change creates the conditions for collective well-being. We need **both** urgently.

How Personal Action Builds Systemic Pressure

Every time you make a conscious choice to bike instead of drive, to write a letter to your representative, to join a protest, or to vote with the planet in mind, you're not acting alone. You are:

- Sending market signals (as a consumer).
- Modelling norms (as a neighbour, parent, or teacher).
- Building momentum (as one part of a growing movement).
- Shifting the story (by redefining what "normal" looks like).

For example:

- When millions of people reduced their use of plastic straws, it gave cities and states leverage to ban single-use plastics.
- As solar panels became more common on rooftops, they became cheaper and more politically viable.
- When households compost, cities take note, and many launch curbside programs as a result.

Personal action isn't the end; it's just the beginning. It lays the emotional, cultural, and political groundwork for a larger transformation.

How Systemic Change Empowers Personal Action

Likewise, when laws, policies, or corporate norms change, individual choices become easier, even automatic.

- When a city bans plastic bags, everyone brings a reusable one.
- When public transit becomes safer and more reliable, more people use it, cutting emissions without needing a speech.
- When governments offer rebates for insulation or electric cars, the products, which were previously unaffordable, become affordable after the rebates.

Systemic change scales up good habits and removes the friction that stops people from doing the right thing. It also protects those who've already been acting sustainably by aligning public infrastructure, funding, and education with their values.

From Action to Advocacy, Finding Your Leverage

Everyone has a different level of influence. Some write policies. Some teach kids. Some organize their street or building. Some simply lead by example.

The sweet spot is finding ways to **align your personal habits with broader goals**, such as:

- Eating plant-rich meals and advocating for healthy, sustainable food programs in schools.
- Riding your bike and helping push for bike lanes and traffic-calming measures.
- Supporting local businesses and helping draft city procurement guidelines that favour green suppliers.

By connecting your values, behaviours, and voice to existing campaigns or political processes, you amplify your impact.

You don't need to be an expert, just engaged.

We Need All Hands, All Levels

The climate and ecological crises require rapid, large-scale transformation. But that doesn't mean your small-scale actions don't matter.

They matter because:

- They build community readiness.
- They increase pressure on institutions.
- They create a culture that supports bold decisions.

Your reusable bag may not save the planet, but it might help normalize a low-waste lifestyle. Your decision to grow vegetables won't end industrial farming. Still, it may influence a neighbour, a child, or a city planner.

In this way, personal change becomes cultural change, and cultural change, in turn, becomes policy change.

So don't choose between the personal and the systemic. Live your values *and* advocate for systems that enable others to do the same.

Forms

Personal + Systemic Change Reflection Page

Use this page to reflect on how your personal choices and systemic efforts work together.

There's no one right path. Your power lies in what you combine and share with others.

1. What are 2 or 3 personal changes I've made (or want to make)?
2. What bigger systems do these actions connect to (e.g, energy, waste, transport, food)?
3. Have I seen any of my habits influence others? If so, how?
4. What's one system-level issue I care about (e.g., clean air, tree protection, composting)?
5. What's one simple way I could take action on that issue beyond my personal habits?
6. What support would I need (tools, teammates, knowledge)?
7. Who could I share this conversation with?

Remember: your actions help shift culture, and culture moves policy.

Change Circle Activity

Draw or write in the circles below. Start with your personal actions, then connect them to community, cultural, and systemic shifts. Use arrows to show how they relate!

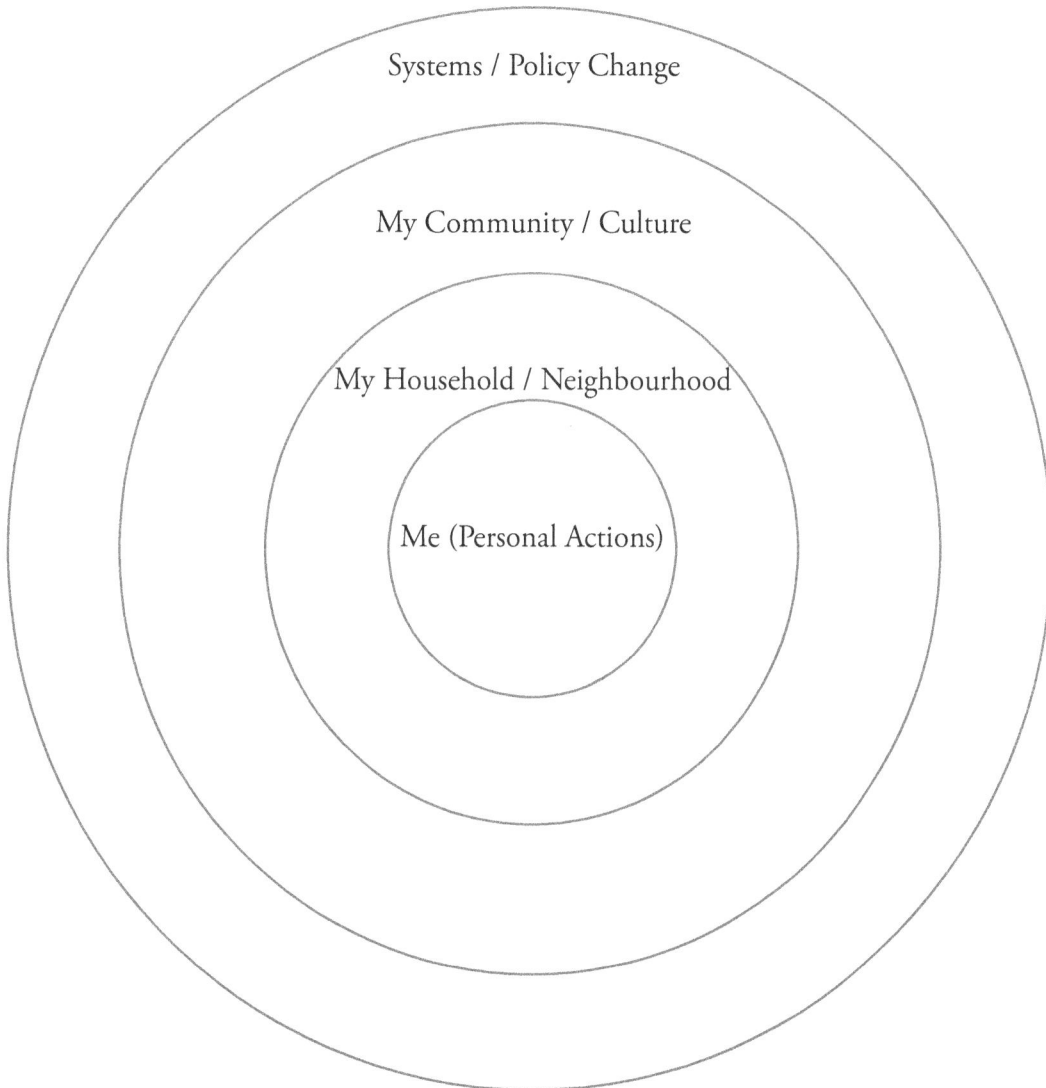

Systems / Policy Change

My Community / Culture

My Household / Neighbourhood

Me (Personal Actions)

Use this map to ask: Where am I strong? Where could I grow? Who can I team up with?

How to Speak Up Without Burning Out

Why Your Voice Matters

You may not be a politician, lobbyist, or professional campaigner. But your voice, as a voter, consumer, parent, student, worker, or neighbour, carries weight.

When citizens speak up, they:

- Show decision-makers what the public cares about.
- Normalize environmental concerns in everyday conversations.
- Shift the culture of workplaces, schools, and communities.
- Make it easier for others to join in.

Speaking up doesn't always mean speeches or protests. It can mean:

- Writing a short letter to your city council.
- Asking your school to start a compost program.
- Posting on social media about local issues.
- Discussing policies with friends and family that matter to them.

It all adds up. Advocacy is a team sport, and your voice may be the one that inspires someone else to take action.

Ways to Speak Up (Big and Small)

There are many levels of advocacy. Start with what feels right for you, and grow from there.

Everyday Actions:

- Sign a petition on an issue you care about.
- Send a respectful message to your local representative.
- Share a resource or success story with your community.

Community Influence:

- Speak during a public comment period or town hall.
- Join a neighbourhood group focused on sustainability.
- Suggest ecological improvements at your workplace or place of worship.

Leadership and Beyond:

- Start a campaign or challenge in your school, union, or association.
- Attend a climate rally or support frontline communities.
- Collaborate with others to draft local policy recommendations.

You don't need to do everything. Select a few actions that align with your energy, comfort level, and available time, and then build upon them.

Avoiding Burnout and Staying Grounded

Sustainability isn't just about the planet; it's about you, too. Staying engaged in the long term means taking care of your energy, emotions, and expectations.

Tips for Sustainable Advocacy

- *Focus on one thing at a time.* You can't fix everything, but you can influence something.
- *Work with others.* Shared purpose eases the burden and amplifies your impact.
- *Celebrate wins.* Even small victories matter; honor them.
- *Take breaks.* Rest is resistance. You're more effective when recharged.
- *Stay connected to what you love.* Whether it's a park, a beach, or your kids' future, keep your "why" close.

Burnout happens when hope fades. Keep hope alive by focusing on relationships, progress, and the joy they bring.

What Helps People Keep Going

From long-time activists to first-time advocates, people tend to stay engaged when they feel:

- *Useful.* They know their effort counts.
- *Connected.* They are part of a community or movement.
- *Skilled.* They feel confident about how to act.
- *Supported.* They know they're not alone.
- *Hopeful.* They can see that change is possible.

Your role doesn't have to be loud or public. It can be steady, behind the scenes, or creative. All kinds of advocacy matter.

Speak Up, Stay True, and Stay in the Game

Environmental advocacy is not about being perfect; it's about being present. It's showing up, again and again, for what you believe in.

Remember:

- Speak clearly but kindly.
- Ask questions and listen deeply.
- Join forces and pass the mic when needed.

And don't forget to celebrate your part. You're not alone, and you're not too small to matter.

Forms

Speaking Up Toolkit

Sample Email to a Representative

Dear [Representative's Name],

As a resident of [your area], I am writing to express my concern about [issue].

I urge you to support action on this issue by [suggest action or bill].

This matters to me because [personal reason(s)].

Thank you for your attention and for serving our community.

Sincerely,

[Your name], [Your City and Postal Code]

Public Speaking Checklist:

- What's the key message I want to share?
- Who am I speaking to, and what do they care about?
- Can I use a short story or a personal example?
- Have I practised aloud once or twice?
- Am I breathing and pausing between points?
- Do I invite others to act or follow up?

I Spoke Up For ...

Clean Air	Wildlife	Future Generations

Water Protection	My Community

Case Study: Youth Climate Strikes Changing National Conversations.

A Movement Ignited by One Voice[34]

In August 2018, a 15-year-old girl sat alone outside the Swedish parliament, holding a handmade sign that read: *"Skolstrejk för klimatet"* (School Strike for the Climate).

That girl was *Greta Thunberg*. At first, she was dismissed as idealistic or extreme. But her quiet determination struck a chord with young people around the world who felt their futures were being ignored.

Within months, her solo protest had evolved into *Fridays for Future*, a global youth-led movement demanding that governments take stronger climate action. What began with a cardboard sign became a ripple that reshaped global discourse.

From Local Action to Global Momentum[35]

By early 2019, students in over *125 countries* were participating in school climate strikes. The largest strike in September 2019 drew an estimated *7.6 million people* to the streets, one of the biggest environmental protests in history.

These youth weren't just skipping class. They were organizing marches, writing open letters, holding town halls, testifying before parliaments, and meeting with world leaders. Many created their own local chapters and coalitions, tailoring their messages to national and cultural contexts.

In Uganda, *Vanessa Nakate* launched strikes to highlight how climate injustice disproportionately affects African countries. In the Philippines, youth called out deadly typhoons and mining. In Canada and Australia, Indigenous youth brought land rights into the centre of the conversation.

The climate strike model evolved into more than a protest; it became a *platform for justice*, demanding urgent, science-based, and inclusive action.

Shifting the Narrative, Not Just the Policies

While many governments have yet to deliver on bold promises, the youth strikes have already succeeded critically: they've **shifted the public conversation**.

Before the strikes:
- Climate change was often seen as a distant or abstract issue.
- Political leaders faced little pressure from younger generations.
- Media coverage was sporadic and technical.

After the strikes:
- Climate change became a moral issue tied to fairness, urgency, and youth rights.
- Words like climate emergency, net zero, and just transition entered daily use.
- Youth voices became regular guests on panels, in documentaries, and in political debates.

Surveys in several countries have shown that public concern about climate change *increased sharply* during and after youth-led strikes, particularly among parents and voters.

What We Can Learn from Youth Advocacy

The youth climate strike movement teaches us several key lessons about modern advocacy:

1. *You don't need permission to begin.* Movements often start with small acts of courage.
2. *Your story is powerful.* Young people spoke authentically about fear, hope, and duty—and the world listened.
3. *Action inspires action.* Seeing one person take a stand can move thousands to follow.
4. *Unity doesn't require uniformity.* Youth coordinated globally while adapting locally.
5. *Moral clarity can cut through noise.* The message was simple: *Listen to the science. Protect our future.*

Youth-led strikes also redefined who gets to speak. No longer is climate conversation reserved for scientists and politicians. It belongs to students, frontline communities, and future generations.

Keep Striking—in Your Own Way

Not everyone can leave school or work to join a protest. But the spirit of the climate strikes: urgency, honesty, and courage, lives on in many forms:

- *Join or support youth-led organizations.*
- *Share their stories and amplify their messages.*
- *Attend a local rally or organize a teach-in.*
- *Push for climate education in your school or community.*
- *Call for inter-generational dialogue in your workplace or government.*

The youth climate strike movement didn't wait for change; it *demanded* it. And in doing so, it reminded the world that we are out of time, but not out of options.

Forms

Youth Climate Strikes: Voices for the Future

What started with one voice became a global movement for justice and action.

"You are never too small to make a difference." — Greta Thunberg

"Our house is on fire and we're striking to put it out." — Greta Thunberg

"We want you to panic. We want you to act." — Greta Thunberg.

Fast Facts:

- First strike: August 2018, Sweden

- 7.6 million joined the Global Climate Strike in 2019

- 125+ countries involved

- Youth from every continent contributed ideas and leadership

What would you strike for?

Strike doesn't always mean skipping school; it means stepping up.

Speak, write, march, organize, support, be the change your future needs.

Youth Climate Strike Action Card

A future I want to help build:

One action I will take this month:

Who I can team up with or support:

A message I want to send to leaders:

This is my climate voice. I choose action, not silence.

Youth Climate Strikes Reflection Worksheet

This page helps you reflect on what the youth climate strikes mean to you and how you might act on your own values, voice, and vision for the future.

1. What emotions do the youth strikes stir in you?

2. What do you think was most powerful about their message or methods?

3. What's one climate or justice issue that affects you or your community?

4. If you could speak to a decision-maker, what would you say?

5. What's one action you feel ready to take—big or small?

Remember: your voice matters. Even small actions grow movements.

What Kinds of Policy Make the Biggest Difference?

Not All Policies Are Equal

When it comes to addressing significant environmental challenges, not all actions carry equal weight. Some policies can shift an entire system, while others, though helpful, may only scratch the surface.

To make meaningful change, we must focus on *high-impact, scalable, and enforceable* solutions.

Let's look at a few examples across sectors:

Issue	Low-Impact Policy	High-Impact Policy
Plastic Waste	Banning plastic straws	Extended producer responsibility (EPR) laws
Transportation	Encouraging carpooling	Zero-emission vehicle mandates
Energy	Promoting energy-efficient light bulbs	Phasing out fossil fuel subsidies
Deforestation	Tree-planting days	Protecting old-growth forests legally
Emissions	Voluntary carbon offsets	Binding emissions reduction targets

Key takeaway: Policies that *alter incentives or rules for industries* tend to have the most significant impact.

The Power of *Upstream* Thinking

Environmental problems often manifest *downstream* as litter in rivers, smog in cities, and rising grocery prices due to drought. But effective policy focuses *upstream*: stopping problems before they spread.

Upstream policies might include:

- Requiring companies to design products for reuse or recycling.
- Setting caps on total greenhouse gas emissions.
- Ensuring new buildings meet green standards.
- Regulating pollutants at the point of extraction, not disposal.

These policies may not seem exciting to the public, but they work quietly and powerfully in the background, shifting the entire system.

Remember: upstream = long-term.

Where Policy Meets People

Systemic policies are vital, but they only work if they are:

- Equitable (fair for all groups)
- Enforced (not just suggestions)
- Understood (so people know how they help)

That's where everyday people come in.

Advocacy helps ensure that:

- Climate and conservation policies aren't watered down by corporate pressure.
- Public funds go toward proven solutions.
- Leaders are held accountable to scientific targets.
- Voices from all communities, especially those historically excluded, are heard.

When citizens understand how policy works and why it matters, they're better equipped to *protect*, *promote*, and *participate*.

Focus Areas for Strong Environmental Policy

Here are six key areas where advocacy and legislation can have powerful ripple effects:

1. *Clean Energy Transition*

   ✉ Investment in renewables, grid upgrades, and energy access for all.

2. *Nature Protection*

   ✉ Legal safeguards for biodiversity, forests, wetlands, and oceans.

3. *Climate Targets & Accountability*

   ✉ Binding emissions limits, carbon pricing, and adaptation strategies.

4. *Circular Economy & Waste*

   ✉ Bans on unnecessary single-use items, product take-back laws.

5. *Green Cities & Transport*

   ✉ Transit infrastructure, walkability, bike lanes, and zoning reform.

6. *Just & Inclusive Solutions*

   ✉ Policies that address both climate and inequality, centering Indigenous rights, worker transition, and youth involvement.

These areas give advocates at all levels, from students to city planners, clear focus points.

Policy Is Not a Magic Wand, But It's a Key Puzzle Piece

Policy can't do everything. It can't replace culture, community, or creativity. However, it provides us with *guardrails*, *incentives*, and a clear *vision*.

Combined with education, innovation, and grassroots action, good policy can:

- Drive large-scale change fast.
- Level the playing field.
- Ensure that progress isn't reversed with a change in leadership.

The challenge? Policy often lags behind science. But, with public pressure and persistent hope, we can close that gap.

You don't need to write the law, just raise your voice for those who do.

Forms

Top 6 Policy Areas for a Healthier Planet

Clean Energy
Invest in renewables, enhance grid infrastructure, and ensure universal access for all.
Nature Protection
Safeguard forests, wetlands, oceans, and biodiversity.
Climate Targets
Set binding emissions caps, carbon pricing, and accountability.
Circular Economy
Reduce waste, ban single-use plastics, and promote reuse.
Green Cities
Build transit, bike lanes, walkability, and eco-friendly zoning.
Justice & Inclusion
Centre equity, indigenous rights, and fair worker transitions.

Advocate for policies that shift systems, not just habits. Big change needs big tools.

Citizen Advocacy Toolkit

Write a letter to local leaders.
Start with who you are and why you care. Keep it short, clear, and solution-focused.

Start a petition
Use a platform like change.org or a local tool. Be specific and include a clear ask.

Speak at a council meeting.
Prepare a 1-to-2-minute message. Share a story, provide data, and request a specific action.

Build a coalition
Team up with local groups, schools, elders, or businesses. Power grows with partnerships.

Spread the word online.
Share posts, use hashtags, and invite others to join. Stay respectful and factual.

My local advocacy goal:

Policy starts with people. Your voice is a lever for change.

How to Engage with Democracy Beyond Election Day

Many people think their role in democracy begins and ends at the ballot box. However, real, lasting environmental progress occurs in the *spaces between elections*: in city halls, through petitions, in public comment periods, and in everyday civic conversations.

Democracy Is More Than Voting

Yes, voting matters. But democracy is not a one-day event; it's an ongoing conversation between the people and their institutions.

Here's how everyday people shape environmental progress between elections:

Action	Impact
Attend a public consultation	Help shape city policies, budgets, or zoning
Speak at a council meeting	Put local sustainability issues on the agenda
Start a petition	Build support and pressure for change
Contact your representative	Let them know what matters to their voters
Join or form a climate caucus	Influence your party or community platform

Each of the actions above adds weight to environmental concerns in decision-making spaces, especially when repeated over time or coordinated with others.

Knowing Who Makes the Rules

If you want to change something, you have to know who's in charge.

Different environmental issues are handled at different levels of government:

Government Level	Handles...
Local (City/Town)	Recycling, transit, parks, development, and tree bylaws
Provincial/State	Education, resource permits, energy, and healthcare
Federal/National	Emissions targets, international treaties, and fisheries

When you know which level handles what, you can *target your advocacy* effectively.

Tip: If you're unsure, ask your local councillor or use a "Who's My Rep?" website.

Staying Loud and Kind

Speaking up matters, but how you speak up also matters. Here are some tips for staying active, confident, and heard:

- *Be respectful, even if you disagree.* This builds credibility and keeps the door open for future opportunities.
- *Tell a short, personal story.* People remember stories more than stats.
- *Be clear about what you're asking for.* "Support X policy," "Vote against Y bill," etc.
- *Repeat your message.* Change takes persistence.
- *Say thank you.* Public service is hard. Gratitude gets noticed.

Remember: You don't need to be an expert to care. Passion, lived experience, and clarity are powerful tools.

Building Civic Habits

Just like composting or biking to work, civic engagement gets easier the more you do it. Here's a simple monthly practice you can try:

The "12-Month Civic Action Calendar"

Month	Action
January	Send a thank-you note to an environmental ally
February	Attend a local climate or sustainability event
March	Write your rep about a current issue
April	Volunteer at a green event for Earth Month
May	Learn about a new environmental policy
June	Organize a small awareness campaign
July	Join a local or youth climate group
August	Submit a letter to the editor
September	Help with a school or community green project
October	Host or join a climate discussion
November	Encourage others to register to vote
December	Reflect, review, and recommit for the year ahead

By year's end, you'll have built a habit of public participation and likely inspired others to do the same.

Forms

Who Represents Me?

Use this worksheet to find and record the names and contact information of your elected representatives at different levels of government.

Local (City or Town Council)

Handles: recycling, parks, zoning, local climate plans

Name:	
Phone or Email:	
Role or Committee:	

Regional or Provincial/State

Handles: education, energy, health, land, and water use

Name:	
Phone or Email:	
Role or Committee:	

National/Federal

Handles: emissions targets, international policy, conservation laws

Name:	
Phone or Email:	
Role or Committee:	

Tip: Use an online 'Who's My Representative?' tool to look up this information.

Making Change from Where You Are

You don't need to be a politician, lawyer, or celebrity to shift systems. Some of the most effective change-makers are ordinary people who start right where they live, work, or study.

Start Where You Stand

It's easy to feel like policy change is far away, in government buildings or corporate offices. However, every job, classroom, community group, or faith circle is part of a larger system.

Here's what *starting where you stand* might look like:

Your Role	Change You Can Make
Student	Propose a school climate action plan
Employee	Suggest a green workplace policy or audit
Parent	Speak up at a school board or PTA meeting
Elder or Retiree	Share lived experience in community consultations
Artist/Storyteller	Create work that inspires and educates
Volunteer/Member	Help local groups draft or promote petitions

Start by asking: "What decisions are being made around me, and who can I talk to?"

Advocacy in Everyday Jobs

Most jobs, even those outside the environmental field, involve systems that matter: food, transportation, buildings, energy, healthcare, and so on.

Examples of *policy-relevant advocacy within workplaces*:

- A nurse pushes for hospital-wide climate and health guidelines.
- A restaurant manager reduces food waste and joins a city food policy council.
- A tech worker advocates for their company to divest from fossil fuels.
- A public librarian hosts a local environmental speaker series.

Small workplace actions can lead to bigger policy shifts, especially when staff organize together.

Policy Doesn't Always Start at the Top

Some of the most significant environmental laws didn't originate with governments. They started with community resolutions, union demands, or student-led proposals.

Consider:

- Anti-idling bylaws proposed by youth councils.
- Campus divestment campaigns that reshaped institutional investment.
- Public transit expansions rooted in neighbourhood petitions.
- Tree protections were added to the city bylaws after community surveys.

These examples demonstrate that local action fosters political will, which, in turn, informs higher-level decisions. One spark can light a fuse.

The "Ripples of Influence" Framework

Below is the Ripples of Influence model:

[You]

✉ Your household

✉ Your friends or coworkers

✉ Your school, workplace, or local government

✉ Regional and national policy

Each outer ripple starts from the centre, and every action has the potential to ripple outward.

Your Voice + Their Power = Policy in Motion

Policy is not about perfect knowledge; it's about values, representation, and the courage to act. When citizens bring personal stakes and grounded stories to the table, decision-makers are more likely to listen.

To keep policy in motion:

- Show up often, not just once.
- Partner with others who share your goals and objectives.
- Frame your cause around justice, science, and shared values.
- Speak as someone with experience, not just an opinion.

Systemic change needs both insiders and outsiders. Find your doorway in and open it for others, too.

Forms

My Influence Circles

Use this page to explore how your actions and voice have a ripple effect.

In each ring, write who you influence and how you can inspire change at home, at work, at school, and beyond.

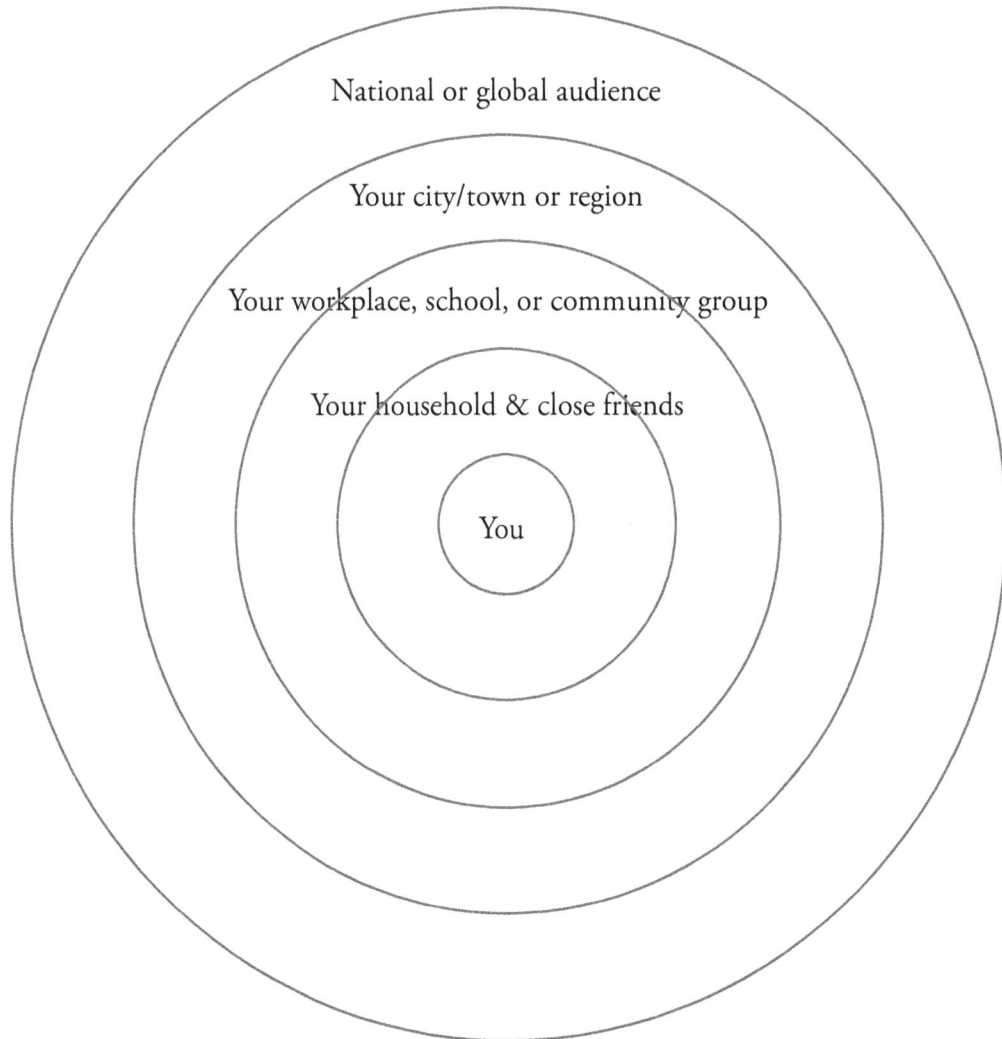

National or global audience

Your city/town or region

Your workplace, school, or community group

Your household & close friends

You

No circle is too small to matter. All influence begins with one voice.

Telling the Story: Advocacy Through Art, Media, and Culture

You don't need a law degree to change minds. Culture, from music to murals, poetry to protest signs, plays a significant role in shaping public understanding and influencing political will.

This section examines how storytelling and creative expression enable environmental advocacy to resonate with hearts, not just policies.

Why Culture Matters

Policy change requires public support. And public support starts with stories that move us.

Culture reaches people where facts alone cannot. Below are some examples:

- A documentary can show what a report cannot.
- A protest song can rally people faster than a press release.
- A mural can turn an abandoned wall into a symbol of resistance.
- A children's book can plant a lifelong seed of stewardship.

Culture doesn't just *reflect* society; it helps *shape* it.

The Power of Storytelling

Every effective movement has a narrative. Think of the following examples:

- Greta Thunberg's school strike—a lone student sitting outside parliament.
- The Standing Rock protests—protecting water, led by Indigenous voices.
- The phrase "There is no Planet B" is now global shorthand for urgency.

Stories help people in the following ways:

- Understand why change is needed.
- See themselves as part of something bigger.
- Remember the cause long after the moment has passed.

Tip: You don't need to invent a story. Just tell your own truth, with honesty and clarity.

Tools for Creative Advocacy

Here are some cultural strategies that amplify environmental messages:

Tool	Impact Example
Public Art	Murals, chalk art, poster campaigns
Film/Video	Documentaries, TikToks, mini-interviews
Music & Performance	Protest songs, street theatre, flash mobs
Writing & Poetry	Op-eds, spoken word, zines, letters
Photography	Before-and-after shots, photo essays
Design	Memes, infographics, eco-logos

Use what you already love, and use it to speak out.

Social Media, Storytelling at Scale

Love it or hate it, social media is today's most powerful storytelling platform.

The following are some tips for using it well:

- *Post with purpose.* Don't just share. Invite action.
- *Use visuals.* Photos, videos, and simple infographics are effective at grabbing attention.
- *Tell short stories.* Use your captions to explain what, why, and how.
- *Be consistent.* One post is a start. Regular sharing builds awareness.
- *Tag and amplify.* Mention organizations, allies, or local reps to connect.

Even a post seen by 20 people can spark 2 conversations, and one real change.

Your Voice, Your Medium

Everyone has a medium, a way they naturally communicate. Yours might be one of the following:

- Writing a letter or blog post.
- Illustrating a scene from your community.
- Recording a short voice message or a song.
- Designing a digital campaign or art piece.

Ask yourself the following questions:

- What issues do I care most about?
- What art or media am I drawn to?
- How can I connect these to inform, inspire, or influence others?

Don't worry about being *good.* Worry about being *honest, clear,* and *persistent.*

Forms

Creative Advocacy Planner

Use this planner to sketch out your creative idea for sharing an environmental message.
You can use writing art, video, music, social media, or any format you like.

What is your message?
Keep it clear and memorable.

Whom do you want to reach?
Consider factors such as age, community, and location.

What medium will you use?
Art, writing, video, music, performance, etc.

Where will it appear?
Online, at school, in a gallery, or at a local park, etc.

Who can help you or collaborate?
Friends, teachers, artists, organizations.

When will you launch or share it?
Pick a date or event.

Creativity changes minds. Tell your story. Share your vision.

Speak Up for the Planet!

Main Message or Slogan:

(Write or draw your key message here)

What You Can Do:

(List actions like reduce waste, vote, plant trees, ...)

More Info / Links / QR Code:

(Optional - add ways to learn more)

This poster was made by:

When You Speak, the System Listens

Systemic change might seem slow, but it always starts with momentum, and momentum begins with someone speaking up.

This chapter has shown that policy isn't just the domain of politicians. It also belongs to the following people:

- People who raise their voices with purpose
- People who connect local stories to global causes
- People who use tools like voting, organizing, and storytelling
- People who keep pushing, even when the path feels long

Your role in environmental policy is not to do it all, but to **do what you can from where you are**, and support others doing the same.

Your Policy Power Summary Table

What You Can Do	Why It Matters
Vote in every election	Elects leaders who shape environmental rules
Write or meet your rep	Reminds them of whom they represent
Join or start campaigns	Builds public pressure and political will
Speak up in your job/school	Connects daily life to big policy ideas
Create or share content	Changes minds, spreads awareness
Support frontline voices	Centres justice, equity, and lived experience

Each of these actions is a form of **citizenship**, not just advocacy.

Journal Prompt:

1. What environmental issue do I care most about right now?
2. Who is currently making the decisions that affect it?
3. What is one small action I can take to influence that system?
4. Who can I team up with or learn from?

Change doesn't require perfection. It requires persistence—and partnership.

A Final Word: You Are Already Part of the Movement

If you've read this far, you've already begun.

Whether your path leads you to the local school board, a neighbourhood petition, or a viral video, you are part of a long and powerful story: people organizing for a livable planet.

Don't wait to be invited. **You belong here.**

Forms

Your Policy Power

Policy change isn't only for politicians. It's powered by people who show up, speak out, and help others do the same. Here's how you can turn your voice into a visible impact:

What You Can Do	Why It Matters
Vote in every election	Elects leaders who shape environmental rules
Write or meet your rep.	Reminds them of whom they represent
Join or start campaigns.	Builds public pressure and political will
Speak up in your workplace or school.	Connects daily life to big policy ideas
Create or share content.	Change minds, spread awareness.
Support frontline voices	Centres justice, equity, and lived experience

Each action is a form of citizenship. You already have the tools; now use them.

Final Reflection: My Policy Power

Take a moment to reflect on your voice, your values, and your next steps.

There are no wrong answers, just honest ones. This page is for you.

1. What environmental issue do I care most about right now?

2. Who is currently making the decisions that affect this issue?

3. What is one small action I can take to influence that system?

4. Who can I team up with or learn from?

5. How will I follow through, and when?

Your reflection can be the beginning of something powerful. Keep going.

Your Stewardship Steps

Reflect

Look around your neighbourhood. What local environmental problem or opportunity do you notice most? Is there litter in a park, an empty lot that could be a garden, or a stream that needs care? Who do you know—neighbours, friends, teachers—who might care too? Seeing your community through a steward's eyes helps you find your place to make a difference together.

Learn

Look up what local groups, schools, or neighbours are already doing to care for the environment. Is there a community garden, a wildlife rescue group, a litter pick-up day, or a neighbourhood tree planting? Follow one on social media or drop by an event to see what they're up to. Learning about local action reveals that real change often begins right next door.

Act

Use your voice for the bigger picture. Sign a petition for a cause you care about. Write or email your local representative to ask what they're doing to protect nature and fight climate change. Discuss with friends why voting with the planet in mind is important. Change happens when citizens speak up and keep speaking up.

Stewardship Journal

What did I learn?

_____

_____

_____

_____

_____

_____

What small action will I try this week?

_____

_____

_____

_____

_____

_____

With whom will I share what I have learned?

_____

_____

_____

_____

_____

Chapter 9:

Global Connections, Shared Responsibilities

Introduction

One Planet, Shared Consequences

- We live on a single, interconnected planet. A river polluted in one country can carry waste to another. Carbon dioxide released in one city changes the climate everywhere. A forest preserved by one community can help cool the entire globe.
- Our actions ripple across borders—for better or worse.
- Whether you're saving energy in Ontario, protecting mangroves in the Philippines, or planting native trees in Kenya, you are part of a web of people working for the same goal: a livable, just, and thriving Earth.
- This is the power and responsibility of **global citizenship**.

The Environmental Justice Lens

- Environmental challenges are not felt equally. While wealthy nations have contributed the most to global warming and waste, low-income communities and countries often suffer the worst effects: droughts, floods, pollution, and food shortages.
- This is where **environmental justice** comes in.
- It reminds us that the climate crisis is also a human rights crisis, that the people with the fewest resources must not be left with the heaviest burdens, and that true solutions uplift **everyone**, not just the privileged.
- **Stewardship is not just individual. It's collective, fair, and global.**

What It Means to Be a Global Steward

A global steward is someone who engages in the following actions:

- Thinks beyond borders
- Acts with solidarity
- Listens to and uplifts frontline communities
- Takes responsibility for their part and shares tools, not just opinions

Global stewards aren't perfect. However, they are diligent in the following activities:

- Staying informed
- Sharing resources and stories
- Supporting fair trade, ethical investments, and global education
- Holding leaders accountable on the world stage

We are many, but we share one home.

Everyday Actions with Global Impact

Here are simple, local actions that connect to global outcomes:

Action You Take	Global Impact
Choose reusables	Cuts plastic waste affecting ocean ecosystems
Eat less meat/dairy	Reduces deforestation and methane emissions
Buy fair trade or local	Supports worker rights and reduces exploitation
Save energy at home	Helps cut global carbon demand
Avoid fast fashion	Lowers pollution and waste in overseas factories
Speak up for climate justice	Supports fair policies across borders

You don't need a passport to make a global difference.

Case Studies: Cross-Border Environmental Cooperation

Case Study 1: The Great Green Wall of Africa[36]

Countries involved: Over 20 African nations

Goal: Combat desertification, restore degraded land, improve food security

The Great Green Wall is one of the most ambitious ecological restoration efforts on Earth. Stretching across the Sahel—from Senegal in the west to Djibouti in the east—this initiative aims to plant trees, rehabilitate land, and bring life back to an area severely affected by climate change and unsustainable farming.

Successes:

- Millions of hectares of land restored
- Over 300,000 jobs created in rural communities
- Food and water security improved in participating regions

Global relevance: Helps sequester carbon, improves resilience, and sets a precedent for cooperation in climate-impacted zones.

Case Study 2: Protecting the Coral Triangle[37]

Countries involved: Indonesia, Malaysia, Papua New Guinea, Philippines, Solomon Islands, East Timor

Goal: Preserve the world's richest marine biodiversity region

The Coral Triangle is known as the *Amazon of the Seas*. It's home to 76% of known coral species and thousands of marine creatures. Coastal nations formed the **Coral Triangle Initiative**, coordinating marine protected areas, sustainable fishing policies, and community engagement.

Highlights:

- Shared conservation science and enforcement tools
- Locally led reef monitoring and marine zoning
- Strong international donor and NGO support

Global relevance: Protects critical biodiversity, fisheries, and coastal economies that feed millions.

Case Study 3: The Amazon Cooperation Treaty Organization (ACTO)[38]

Countries involved: Brazil, Bolivia, Colombia, Ecuador, Guyana, Peru, Suriname, Venezuela

Goal: Coordinate Amazon rainforest conservation and sustainable development

The Amazon rainforest spans eight countries and is vital for global carbon storage and climate regulation. ACTO facilitates information-sharing, research, joint planning, and responses to threats like deforestation and illegal mining.

Notable efforts:

- Trans-boundary protected area planning
- Joint water resource management in the Amazon Basin
- Indigenous community participation in governance

Global relevance: Preserves one of the planet's largest carbon sinks and key biodiversity zones.

Case Study 4: Arctic Council Environmental Protection[39]

Countries involved: Canada, Denmark (Greenland), Finland, Iceland, Norway, Russia, Sweden, United States

Goal: Protect Arctic ecosystems while managing increasing human activity

Despite political tensions, Arctic nations have collaborated for decades through the Arctic Council, focusing on pollution control, sustainable development, and climate monitoring.

Collaborative wins:

- Shared climate impact data and research
- Regulations on shipping and oil spill response
- Indigenous knowledge integration in decision-making

Global relevance: Offers a model of peaceful, science-based cooperation in a geopolitically sensitive region.

Forms

Reader Pledge: I Am a Global Steward

I recognize that my choices affect not just my neighbourhood but my planet.

I pledge to act with awareness, compassion, and justice across borders, oceans, and cultures.

I will:

- Make ethical, sustainable choices wherever possible
- Learn from people and communities beyond my own
- Uplift frontline voices and solutions
- Use my privileges to protect and preserve our shared future

We are many, but we share one home.

Signed: _____

Date: _____

Global Action Tracker

Use this tracker to reflect on how your daily choices connect to global issues.

Each row helps you connect a local action to a wider impact on people and planet.

Local Action	Global Connection	People Affected	Next Steps

Even small local actions can create global ripples of change.

Who Made This? Supply Chain Management

Pick a common item (e.g., T-shirt, phone, chocolate bar). Use this sheet to map its journey from resource extraction to your hands. Learn where each part comes from and who is involved.

My Item:
What did you choose to research?

Raw Materials:
Where were the ingredients or parts sourced? (e.g., cotton, minerals, cacao)

Manufacturing:
Where was it assembled or processed? Who made it?

Transportation:
How far did it travel to reach you? What modes of transport were used?

Labour Conditions:
Were workers treated fairly? Can you find certifications (e.g., fair trade)?

End of Life:
What happens when you're done with it? Is it recyclable, reusable, or landfill-bound?

Knowing the story behind our stuff is the first step toward ethical choices.

Prominent Advocates

"Twenty-five years ago, people could be excused for not knowing much, or doing much, about climate change. Today we have no excuse."

Desmond Tutu (South Africa)

"Rainforests are the lungs of our planet, providing us with oxygen, purifying the air we breathe, and regulating the climate."

Rainforest Foundation

"You are never too small to make a difference."

Greta Thunberg (Sweden)

"For centuries, indigenous peoples have protected the environment, which provides them with food, medicine, and so much more. Now it's time to protect their unique traditional knowledge that can bring concrete solutions to implement sustainable development goals and fight climate change."

Hindou Oumarou Ibrahim (Chad)

5 Global Environmental Challenges & Shared Solutions

Challenge: Climate Change

Shared Solution: Cut emissions, invest in renewables, and protect carbon sinks, like forests.

Challenge: Plastic Pollution

Shared Solution: Ban single-use plastics, improve recycling, and support circular economies.

Challenge: Biodiversity Loss

Shared Solution: Protect habitats, reduce pesticide use, and support indigenous conservation.

Challenge: Water Scarcity

Shared Solution: Conserve water, treat wastewater, and ensure fair access for all communities.

Challenge: Deforestation

Shared Solution: Enforce forest protection, support sustainable agriculture, and reforest degraded land.

Every region has a role. Every solution needs many hands.

Reader Pledge: One Planet, Shared Care

I understand that environmental responsibility does not stop at my doorstep. I pledge to act with awareness that my decisions affect ecosystems, workers, and communities far beyond my own.

I will:

- Choose products that are fair, sustainable, and ethical
- Reduce waste and emissions that impact people worldwide
- Uplift the voices of frontline and Indigenous communities
- Learn from global movements and share what I discover
- Support policies that protect our shared home for everyone

We are many. We are connected. We are responsible.

Sign: _____

Date: _____

Prominent Advocates Match Game

Match each quote to the correct voice or region. Use this activity to explore diverse perspectives in global environmental movements. Draw lines or write letters to connect quotes to the right person or region.

A. "You are never too small to make a difference."	1. Desmond Tutu - South Africa
B. "For centuries, indigenous peoples have protected the environment, which provides them with food, medicine, and so much more. Now it's time to protect their unique traditional knowledge that can bring concrete solutions to implement sustainable development goals and fight climate change."	2. Rainforest Foundation
C. "Rainforests are the lungs of our planet, providing us with oxygen, purifying the air we breathe, and regulating the climate."	3. Greta Thunberg - Sweden
D. "Twenty-five years ago, people could be excused for not knowing much, or doing much, about climate change. Today, we have no excuse."	4. Hindou Oumarou Ibrahim - Chad

How many did you know already? Which voices were new to you?

Supply Chain Sorting Cards

Cotton Field (India)	Textile Mill (Bangladesh)	Sewing Factory (Vietnam)
Cargo Ship (Indian Ocean)	Retail Store (Canada)	Plastic Packaging Plant (China)
Cacao Farm (Ghana)	Smart phone Assembly (China)	E-Waste Site (Nigeria)

My Global Footprint Quiz

Answer the questions below to reflect on how your everyday choices affect people and ecosystems around the world. Circle or check your answers. Be honest; this quiz is about awareness, not judgment!

1. How often do you buy items made in other countries?

☐ Never	☐ Sometimes	☐ Often	☐ Always

2. Do you check where your food, clothes, or electronics come from?

☐ Never	☐ Sometimes	☐ Often	☐ Always

3. Do you try to avoid products with excessive packaging?

☐ Never	☐ Sometimes	☐ Often	☐ Always

4. Do you look for fair trade or ethical certifications when you shop?

☐ Never	☐ Sometimes	☐ Often	☐ Always

5. How often do you think about how your waste is handled after disposal?

☐ Never	☐ Sometimes	☐ Often	☐ Always

6. Have you ever learned or shared a story from a community different from yours?

☐ Never	☐ Sometimes	☐ Often	☐ Always

7. Do you believe your small actions can contribute to global change?

☐ Never	☐ Sometimes	☐ Often	☐ Always

What's one change you could make to shrink your global footprint?

Awareness is the first step towards transformation.

Fair Trade Hunt

Go on a fair trade label scavenger hunt! Look in your home, local store, or online.
Find as many fair trade or ethical certifications as you can. List the product and the label.

Product	Fair Trade or Ethical Label	Where I Found It

Fair trade means fairness for people and the planet. Every label tells a story.

BONUS: Which label or product surprised you and why?

Reflection Prompt: How Am I Connected?

Encourage learners or readers to journal or discuss:

1. What's one item I use every day that connects me to someone else's labour or land?
2. How could I reduce harm or increase fairness in that connection?
3. Have I ever learned something valuable from a culture or region different from mine? What was it?
4. What does being a *global citizen* mean to me, really?

Your Stewardship Steps

Reflect

Ask yourself: What global environmental problem concerns me most? Where do I see inequality in its effects and in the solutions?

Learn

Understand how environmental impacts cross borders. Explore how climate change, pollution, biodiversity loss, and deforestation are interconnected worldwide, and how justice is tied to environmental solutions.

Act

- Choose one purchase each week that supports global equity (e.g., fair trade, sustainable seafood, or local goods).
- Reduce your consumption of goods made in polluting or exploitative conditions.
- Support international climate justice efforts through donations or campaign visibility.
- Learn from communities on the front lines and amplify their voices.

Share

- Host a discussion about global responsibility—at school, online, or in your faith/community group.
- Post a fact, story, or infographic that connects local actions to global effects.
- Translate what you learn into multiple languages or share content from diverse voices.

Stewardship Journal

What did I learn?

_____

_____

_____

_____

_____

_____

What small action will I try this week?

_____

_____

_____

_____

_____

_____

With whom will I share what I have learned?

_____

_____

_____

_____

_____

_____

<h1>Chapter 10:</h1>

<h1>The Road Ahead</h1>

Introduction: A Fork in the Path

We live at a turning point in human history. The challenges we face—climate disruption, biodiversity collapse, pollution, injustice—are vast. But so is our power to meet them. Across the globe, people are standing up not just to protect what we have, but to reimagine what's possible.

This chapter is not about wrapping things up neatly; it's about equipping you to carry the torch forward. We'll explore what long-term stewardship looks like in a rapidly changing world, how to stay engaged without burning out, and how to find your unique role in building a more just, sustainable future.

What Kind of World Are We Creating?

The future is not written yet. It is shaped every day by our choices: what we consume, how we vote, who we support, and how we treat each other and the Earth.

There are two paths ahead:

- **Business as usual**, where short-term profit continues to outweigh planetary boundaries.
- **Transformation**, where we rise to the challenge with innovation, empathy, and collective action.

Young people, communities, Indigenous leaders, scientists, farmers, artists, policymakers, and citizens—you—are all part of this unfolding story. Every person matters. Every action counts.

From Sustainability to Regeneration

In this final chapter, we move from sustaining what remains to **regenerating what we've lost**. This includes the following:

- Healing degraded ecosystems
- Creating circular economies that eliminate waste
- Restoring climate balance through nature and technology
- Building cultures of cooperation, fairness, and compassion

Regeneration means more than simply doing less harm. It means giving back more than we take.

What You'll Find in This Chapter:

- A visual timeline of possible futures (hopeful and cautionary)
- Tools for long-term engagement and avoiding burnout
- A guide to building your personal sustainability plan
- Reflections from youth and elders on legacy and responsibility
- Final Stewardship Steps, reader pledges, and inspiration for the road ahead

Timeline of Possibilities: What the Next 25 Years Could Look Like

This is not a prediction; it's a fork in the road.

We are standing on the edge of critical decisions. The next 25 years will define the stability of our climate, the health of our ecosystems, and the future well-being of billions of people. Below is a visual thought-experiment: two diverging timelines: one if we stay the course, the other if we shift direction boldly and collectively.

Timeline 1: **Business as Usual** *(2025–2050)*

2025–2030

- Global emissions continue to rise, despite international pledges.
- Biodiversity loss accelerates: more species go extinct, many permanently.
- Plastics production doubles. Waste management systems buckle under pressure.
- Droughts and heatwaves hit new records. Poorer nations suffer the worst.
- Environmental defenders face increasing violence and suppression.

2030–2040

- Coastal cities begin large-scale relocation due to sea level rise and storms.
- Global inequality worsens as climate disasters displace millions.
- Some governments begin rationing water and energy during the summer months.
- Coral reefs largely collapse; fish stocks plummet in tropical regions.
- Political polarization grows as climate refugees face rising hostility.

2040–2050

- Carbon removal becomes an emergency priority, but it's too late for many systems.
- Climate tipping points—ice melt, rainforest dieback—pass thresholds.
- Global food insecurity drives conflict in vulnerable regions.
- Young generations inherit fragile democracies and fractured ecosystems.
- The Earth is 2.5–3°C warmer. The most vulnerable are left behind.

Timeline 2: A Regenerative Future (2025–2050)

2025–2030

- Fossil fuel subsidies end. Clean energy jobs grow rapidly across the globe.
- Cities invest in walkable designs, efficient housing, and green infrastructure.
- Indigenous communities co-lead land restoration and conservation efforts.
- Climate education becomes core to public school curricula worldwide.
- Plastic bans expand, and circular economies emerge at local and national levels.

2030–2040

- Rewilding projects restore major ecosystems—wetlands, forests, and grasslands.
- Carbon drawdown through agroforestry and soil regeneration expands.
- Fair trade and ethical sourcing become mainstream, not niche.
- Multinational youth coalitions influence global policy.
- Pollinators and native species recover in many regions.

2040–2050

- Global carbon emissions hit net zero, then go negative.
- Cooperative global governance is strengthened by shared climate wins.
- Refugees from climate-affected zones are embraced, not criminalized.
- Renewable energy powers 90% of the grid; fossil fuels become legacy tech.
- Cultural narratives shift: regeneration, equity, and interdependence become guiding values.

Why It Matters

The future is not one or the other; it will likely be a mixture. But **which path becomes dominant depends on what we do today**. Every policy, every purchase, every act of care or advocacy is part of that timeline.

We all have a part to play, not in fear, but in *determination*.

Forms

Tools for Reflection and Commitment

Use the following prompts to reflect on your journey as a steward of the earth. This is a space to clarify your values, identify your strengths, and commit to sustainable action in your life and community.

What moments in this journey impacted you most?

What skills, interests, or experiences do you bring to the movement?

What issue do you feel most called to work on (e.g., climate, waste, wildlife, justice)?

What is one habit or mindset shift you want to carry forward?

Who will support you, and who can you support in return?

What is one thing you'll do in the next 30 days to start or deepen your commitment?

"Hope is a discipline." — *Mariame Kaba*

The Long Game: How to Stay Engaged Without Burning Out

Activism is Not a Sprint

Fighting for a better world is a lifelong effort, not a weekend project. Like the Earth itself, we have rhythms. We have limits. And if we don't take care of ourselves, we risk exhaustion, frustration, or even giving up altogether.

This section is about longevity. It's about how to keep going, not out of guilt or panic, but out of purpose, connection, and joy.

What Burnout Can Look Like

- You feel overwhelmed or paralyzed by the scale of the problems.
- You feel guilty whenever you relax or make an imperfect choice.
- You become disconnected from joy, creativity, or your community.
- You stop engaging because you feel like nothing you do matters.

These symptoms are quite *common*. And they can be addressed with care.

Tools for Resilience

1. Know Your Role

You don't have to do everything. Are you an educator? Organizer? Caregiver? Storyteller? Researcher? Artist? Policy nerd? Know your role and embrace it fully.

2. Find Your People

Sustainable movements are built on relationships. Join a community, online or off, where you can share effort, encouragement, and ideas.

3. Balance Action with Rest

Rest is part of the work. Make time for nature walks, creativity, or simply doing nothing. Burnout helps no one.

4. Let Go of Perfectionism

You will make mistakes. You will learn. The goal isn't purity; it's participation and progress.

5. Reconnect with Meaning

Return to the *why*. Whether it's the children in your life, your faith, your love for forests, or your belief in justice, anchor yourself.

6. Celebrate Small Wins

Notice every reusable item, every tree planted, every petition signed, every conversation sparked. These add up.

What Keeps People Going

From interviews and reflections with activists, Indigenous leaders, students, and scientists, the following common themes emerge:

- Love for land and community
- A sense of spiritual or ancestral duty
- Collective joy and shared meals
- Art and music as fuel
- Faith that the future is worth fighting for—even if you don't see the end result

A Mantra to Remember

"We are not required to complete the work, but neither are we free to abandon it."

— *Adapted from Jewish tradition*

You are part of something bigger than yourself. Take breaks when you need them. Pass the baton. Come back when you're ready.

This is the long game.

Forms

Resilience Toolkit

Use this worksheet to build your personal toolkit for staying engaged without burning out. Identify what fuels you, how you restore your energy, and who you can lean on.

What inspires you to stay involved in this work?

What activities bring you peace or joy when you feel overwhelmed?

Who are your support people or communities?

What creative outlets help you express or release your emotions?

What boundaries do you need to protect your well-being?

What is one small thing you can commit to this month that nourishes your resilience?

"Caring for myself is not self-indulgence. It is self-preservation." — Audre Lorde

Know Your Role: Self-Inventory

Everyone has a unique part to play in building a more just, sustainable world. Use this self-inventory to reflect on your strengths, passions, and natural tendencies.

What do people often come to you for help with?

What subjects or issues do you enjoy learning or teaching about?

Do you feel most energized when you're:

- Creating something new (art, projects, solutions)
- Supporting others or caring for your community
- Organizing details, events, or systems
- Researching or investigating facts and trends
- Speaking out, debating, or advocating

Based on your answers, which role(s) do you most relate to?

☐ Storyteller	☐ Educator	☐ Organizer	☐ Healer
☐ Researcher	☐ Advocate	☐ Connector	☐ Creator

How can this role/these roles guide your next steps in sustainability or advocacy?

You don't have to do everything. Just do your part, and do it well.

The Long Game: How to Stay Engaged Without Burning Out

This is a Marathon, Not a Sprint

Caring for the Earth and advocating for change is deeply meaningful work. But it can also be exhausting, frustrating, and overwhelming, especially when progress feels slow or setbacks come hard. The truth is: *you can't pour out of an empty cup.*

To create lasting change, we need not just short bursts of action but a way to *sustain ourselves and each other* for the long haul. This section offers practical wisdom for building long-term resilience.

What Burnout Looks Like

You may be experiencing burnout if you:

- Feel hopeless or powerless in the face of global problems
- Stop engaging because it feels like nothing matters
- Push yourself so hard that you forget to rest
- Lose joy in things that used to inspire you
- Feel guilty when you're not '*doing enough*'

These are normal responses to big emotions and big challenges. You're not alone.

Build Your Resilience Toolkit

Here are some strategies to keep your fire lit without burning out:

1. Know Your Role

You don't have to fix everything. Instead, focus on what you're good at and what brings you alive. Are you a researcher? An organizer? A storyteller? A caregiver? A strategist? A hands-on doer? Lean into your strengths.

2. Embrace Community

Burnout thrives in isolation. Join or create communities where you can vent, laugh, learn, rest, and recharge with others. You don't have to carry this work alone.

3. Rest Is Revolutionary

Rest isn't selfish; it's survival. Nature itself has seasons of rest and renewal. Protect your energy the way you protect the planet.

4. Let Go of Perfection

There's no such thing as a *perfect environmentalist*. We all make mistakes. We all grow. Aim for *progress*, not perfection.

5. Celebrate the Small Wins

Every single positive step matters. A planted seed, a reused bag, a shared article, a single voice raised—they all add up.

6. Reconnect with Purpose

When you feel adrift, return to your *why*. It might be your children, your ancestors, your faith, or your love of the ocean. Keep that anchor close.

Real-World Resilience in Action

Activists, elders, and changemakers around the world stay strong by engaging in the following activities:

* Practising spiritual or cultural rituals
* Sharing food and songs with their communities
* Creating art, stories, and joy
* Taking turns and passing the baton
* Stepping back when needed—and always returning

As author and activist Mariame Kaba says, *"Hope is a discipline."*

You don't have to feel hopeful every moment. But you can choose to keep going anyway.

Reflection Prompt

What are three things you can do to support your long-term engagement with this work?

Write them down. Post them somewhere you'll see. Share them with a friend.

Forms

Personal Sustainability Plan

Use this worksheet to design your plan for living in alignment with your values.
These small, consistent steps add up to meaningful impact—start where you are.

Knowledge & Learning - What topics do you want to learn more about this year?

At Home - What is one habit you can change to reduce your footprint at home?

Food & Consumption - How can you make your meals or shopping more sustainable?

Travel & Commuting - What lower-impact transport options can you commit to?

Advocacy & Community - How will you raise your voice or get involved this season?

Well-Being & Balance - What helps you stay emotionally and physically strong?

"Start where you are. Use what you have. Do what you can." — *Arthur Ashe*

Final Stewardship Pledge

This is your moment to commit; not to perfection, but to participation; not to guilt, but to growth. Use this pledge to express your values and your next steps as an Earth Steward.

- I pledge to act with care, courage, and curiosity.
- I will protect the Earth, not because I must, but because I can and because I love it.
- I will use my voice, my choices, and my skills to help build a just and regenerative world.
- I know I won't do it alone. I will stay connected, stay kind, and keep learning.
- This is not the end. It is the beginning of my stewardship journey.

Name: _____

Date: _____

"We are the ancestors of the future. What we do now matters."

Your Personal Stewardship Plan

My Stewardship Vision:

Write 1–2 sentences about the kind of steward you want to be.

Example: "I want to protect my local environment and make choices that help people and nature thrive, both close to home and around the world."

1 - My Top 3 Stewardship Goals

Goal	Why It Matters to Me	My First Small Step	By When?

2 - People or Groups I'll Connect With

-
-
-

(Example: Local community garden group, online climate action network, friends/family to share ideas.)

3 - How I'll Track My Progress

- I will check in on my goals every _____ (week/month).
- I'll celebrate small wins by _____.
- If I get stuck or discouraged, I'll remind myself _____.

Example: A Filled-In Stewardship Plan

My Stewardship Vision:

"I want to reduce my waste, support my local ecosystem, and inspire my kids to care for the planet."

Goal	Why It Matters to Me	My First Small Step	By When?
1. Cut single-use plastics by 50%	To protect ocean life and reduce waste in my community.	Buy reusable produce bags and water bottles.	This month
2. Start composting food scraps	To enrich my garden soil and keep food waste out of the landfill.	Research a simple backyard compost bin.	Next month
3. Join a local tree-planting event	To improve air quality and green my neighbourhood.	Sign up for the city's volunteer planting day.	This season

People or Groups I'll Connect With:

- Local zero-waste community group
- My kids' school eco club
- Friends to do a monthly clean-up

How I'll Track My Progress:

- I will check in on my goals every month.
- I'll celebrate small wins by having a family picnic in the park.
- If I get stuck or discouraged, I'll remind myself it's not about perfection; it's about progress.

A Quick Stewardship Pledge

"I pledge to do my part as a steward of the Earth to keep learning, make thoughtful choices, and inspire others to care for our shared home."

Ways to Connect with Like-Minded Groups Globally

You're not alone, and you were never meant to be.

Whether you're just beginning your sustainability journey or looking to deepen your impact, connecting with others is one of the most powerful things you can do. Around the world, people are planting trees, cleaning rivers, teaching kids, marching for justice, and rewriting the rules of the economy. There's a place for you in this movement.

Below are some starting points to help you find your community:

Where to Look

Environmental Networks and NGOs

- *350.org* – global grassroots climate movement
- *Greenpeace* – an international campaigning organization
- *Friends of the Earth* – international federation of environmental groups
- *EarthDay.org* – join campaigns and local cleanups worldwide
- *World Wildlife Fund (WWF)* – biodiversity and conservation projects

Youth and Student Movements

- *Fridays for Future* – student-led climate action
- *Youth4Nature* – storytelling and policy influence
- *Roots & Shoots (Jane Goodall Institute)* – youth-driven community action
- *Teach the Future* – sustainability in education advocacy
- *Global Youth Biodiversity Network (GYBN)*

Justice-Based and Indigenous-Led Groups

- *Indigenous Climate Action (Canada)*
- *Amazon Watch*
- *Global Greengrants Fund* – supports local justice initiatives
- *Climate Justice Alliance (USA)*
- *Survival International* – protecting Indigenous rights and land

Skills-Based Platforms

- *Transition Network* – local resilience and re-skilling communities
- *Extinction Rebellion* – nonviolent civil disobedience for systems change
- *Open Source Ecology* – collaborative tech for sustainable living
- *Precious Plastic* – open-source tools for plastic recycling

Digital Ways to Get Involved

Join a global event or campaign

- Earth Hour
- World Cleanup Day
- Climate Strikes
- Plastic-Free July
- ClimateAction.tech

Sign up for online trainings or action hubs

- Learn organizing
- Storytelling
- Lobbying
- Climate science basics.

Follow hashtags and connect on social media

- #ClimateAction
- #YouthForNature
- #JustTransition
- #EcoJustice

Join virtual meetups or forums

- GoodGrub
- Work on Climate
- Discord servers for sustainability topics

What You Can Do With Others

- Join or start a local chapter of a global group
- Organize a screening, workshop, or clean-up day
- Volunteer remotely for research, translation, or digital work
- Collaborate on art, writing, or science with global peers
- Exchange climate stories across borders

Tip: Look Close, Think Big

Not every solution is international; some of the best ones are local. Use these global networks to learn, be inspired, and build bridges, but also take what you learn *home.*

Forms

Global Ally Outreach Card

Use this card to reach out to a group, project, or person making a difference around the world. Whether you're writing to thank them, ask a question, or start a collaboration. This is how we grow connections.

Who I am reaching out to (name/group):

Where they are located (region/country):

What I admire or want to learn from them:

My message of outreach (write, draw, or draft):

How I will contact or share this message:

"Connection is how movements grow. Reach out; you might start something beautiful."

Start Your Own Group: Planning Sheet

Use this sheet to brainstorm and organize the steps to launch a new local or virtual action group. Whether it's a school club, a neighbourhood network, or an online collective, this is your roadmap.

Group Name or Idea:

Purpose or Mission (why does it exist?):

Who is involved (members, partners, advisors):

Where will it meet or operate (in person or online):

First 3 steps I will take:

How will we share what we do (outreach/communication):

What does success look like in 6 months:

"If it doesn't exist yet, maybe you're the one meant to create it."

Books, Podcasts, and Documentaries That Inspire

The environmental movement is filled with stories of resistance and renewal, science and spirit, heartbreak and hope. The titles below are a mix of classics, fresh voices, and hidden gems to keep you learning, reflecting, and inspired long after you finish this guidebook.

Books That Inform and Uplift

- **Braiding Sweetgrass** by Robin Wall Kimmerer

A poetic blend of Indigenous knowledge, botany, and gratitude for the Earth.

- **This Changes Everything** by Naomi Klein

A powerful analysis of capitalism and climate change, and a call to bold systems change.

- **No One Is Too Small to Make a Difference** by Greta Thunberg

Short, fiery speeches by the young climate activist who sparked a global movement.

- **The Future We Choose** by Christiana Figueres and Tom Rivett-Carnac

Optimistic, practical paths forward from two architects of the Paris Agreement.

- **All We Can Save** edited by Ayana Elizabeth Johnson & Katharine K. Wilkinson

Essays, poetry, and art from diverse women leading the climate movement.

- **Drawdown** edited by Paul Hawken

A ranked list of 100 science-backed solutions to reverse global warming.

Podcasts Worth Your Ears

- **How to Save a Planet** (Gimlet)

Smart, funny, hopeful climate stories and solutions.

- **Outrage + Optimism**

Behind-the-scenes perspectives from top climate negotiators and global changemakers.

- **Drilled**

Investigative reporting on fossil fuel influence, climate denial, and accountability.

- **For the Wild**

Deep ecology, Indigenous wisdom, and rewilding conversations.

- **The Yikes Podcast**

Two young activists explore social and climate justice in an accessible, joyful way.

- **Green Dreamer**

Regenerative thinking and decolonizing perspectives on sustainability.

Documentaries That Stay With You

- **2040** (dir. Damon Gameau)

A hopeful look at what the future could be if we adopted existing climate solutions now.

- **The True Cost**

An eye-opening exposé of the human and environmental toll of fast fashion.

- **Our Planet** (Netflix/YouTube)

A visually stunning David Attenborough series showing the beauty and fragility of life on Earth.

- **My Octopus Teacher** (Netflix)

An unlikely friendship with a wild octopus sparks awe, healing, and ecological insight.

- **Before the Flood** (dir. Fisher Stevens, Leonardo DiCaprio)

A global journey through climate science, politics, and activism.

- **Gather** (Netflix)

An uplifting look at Indigenous food sovereignty and resilience across Turtle Island.

Tip: *Make it a shared experience!*

Start a community *eco-book club*, organize a *screening and discussion night*, or recommend a podcast to a friend. Learning together builds stronger movements.

Forms

Watch-and-Discuss Worksheet

Use this worksheet to reflect on a documentary or short video you have watched.
Great for classrooms, family nights, or community screenings.

Title of the documentary/video:

What stood out to you most:

What you learned or feel inspired by:

What questions you still have:

How this connects to your life or community:

One thing you'd like to discuss or share with others:

One action you'll take as a result:

Stories spark change—if we talk about them.

Take Notes and Take Action

Use this page to record insights from what you're reading, watching, or hearing.
Then turn those insights into personal or collective action.

Key insight or quote:

What it made me think or feel:

How it connects to my life or values:

What I want to remember or research more:

Action I will take or talk about:

By when:

Small insights can lead to big changes—if we act on them.

Reading Reflection Sheet

Use this sheet to reflect on any article, book chapter, or story you read.
Great for school use, book clubs, or personal learning.

Title and author:

```


```

What it was about (summary in 2-3 sentences):

```



```

One big idea or insight I took away:

```



```

How it connects to something I already know or believe:

```



```

A quote or moment that stayed with me:

```



```

One thing I want to do or change after reading:

```



```

Notes or reactions:

```



```

"Reading does not change the world, but readers do."

Handy Apps for Earth-Friendly Living

Technology can be a powerful ally in your sustainability journey. Whether you're tracking your carbon footprint, choosing ethical products, or joining local actions, there are apps designed to make doing the right thing a little easier.

Here are some tried-and-true options to explore:

Track Your Carbon Footprint

- **Wren**

Website that enables you to calculate and then offset your carbon footprint.

- **Earth Hero**

Offers a personalized plan to reduce your carbon footprint, with tips and action steps tailored to your lifestyle.

Choose More Ethical and Sustainable Products

- **Good On You**

A fashion rating app that scores clothing brands on labour rights, environmental impact, and animal welfare.

- **EWG Healthy Living**

Scan barcodes on food and personal care products to check for toxins and health hazards. Great for household decision-making.

- **DoneGood**

Curates ethical alternatives for fashion, gifts, and home products, plus discounts for fair trade and BIPOC-owned brands.

Connect with Causes and Local Actions

- **EcoHero**

Challenges, action logs, and community features to build eco-friendly habits and join local efforts.

- **iNaturalist**

Contribute to citizen science by photographing plants and wildlife. Created by the California Academy of Sciences and National Geographic.

- **Buycott**

Scan products to see company affiliations and supply chain ethics. Create campaigns and support aligned businesses.

- **Giki Zero** (UK-based, globally relevant)

Carbon footprint tracker and planner with clear sustainability goals and team tools.

Tips for Using These Tools

- **Start with one app that matches your goals**, like tracking emissions or finding better products.

- **Turn notifications on (or off)** based on your style. Some people love nudges, others prefer quiet reminders.

- **Make it social.** Compete with friends, join teams, or share progress to boost motivation.

- **Delete the guilt.** These are tools for learning and improvement, not perfection.

Eco App Comparison Chart

App	Purpose	Best For
Wren	Carbon offset & tracking	Offsetting with visuals
Earth Hero	Carbon reduction planning	Step-by-step reductions
Good On You	Ethical fashion ratings	Shopping better clothes
EWG Healthy Living	Scan products for toxins	Everyday product safety
DoneGood	Shop ethical products	Conscientious online shopping
EcoHero	Eco habits and local actions	Action-based challenges
iNaturalist	Citizen science ID and logging	Nature observation & ID
Buycott	Ethical consumer campaigns	Scanning & campaign creation
Giki Zero	Carbon footprint and lifestyle	Planning lifestyle changes

Forms

Eco-Tech Scavenger Hunt

Explore apps and digital tools that support sustainable living. Can you find and try each of these?

Use the checklist below to track your discoveries, rate your experiences, and share what you learned!

App/Tool	Found it?	Tried it?	Rating (1-5)	What I Learned
Good On You				
Earth Hero				
EWG Healthy Living				
DoneGood				
iNaturalist				
Buycott				
Wren				
EcoHero				
Giki Zero				

Try them solo or make it a class challenge!

Your Stewardship Steps

Reflect

Look back at the goals you've thought about while reading this book. Which one feels the most doable right now? Which one excites you the most? Which one could you share with family or friends to make it stick? Taking a moment to pause and celebrate your motivation is the first step toward making your plan real.

Learn

Find one or two tools, apps, or websites that can help you stick with your new stewardship habits. This might be a carbon footprint calculator, a local waste sorting guide, or a habit tracker to remind you of your goals. You could even look for an online community or social media group where people share tips and celebrate progress together, so you feel supported every step of the way.

Act

Commit to one low-carbon switch this month. Ride your bike or take public transport once instead of driving. Cook one extra plant-based meal each week. Talk to your workplace or school about cutting energy waste. Simple steps help shrink your carbon footprint and show others how easy climate stewardship can be.

Stewardship Journal

What did I learn?

_____

_____

_____

_____

_____

What small action will I try this week?

_____

_____

_____

_____

_____

With whom will I share what I have learned?

_____

_____

_____

_____

_____

Conclusion: One Earth, One Chance

A Reminder: Stewardship Is Not About Guilt; it's About Care

In a world grappling with environmental breakdowns, it's easy to fall into guilt or despair. We see the plastic in oceans, the smoke in skies, and the shrinking forests, and wonder if our small actions even matter. But here's the truth: environmental stewardship is not about being perfect. It's not about shame. It's about showing up with care.

Stewardship means tending to something you value. Just as you'd care for a friend who's hurting or a garden that needs water, being a steward of the Earth is about practising love, not punishment. Guilt can paralyze us. But care; care moves us to act, to learn, to reach out, to do better bit by bit.

You are not alone in this work. Across the planet, people are planting seeds, literally and figuratively. They are restoring mangroves, mentoring young leaders, challenging policies, mending clothes, and teaching others. Each act, no matter how small, is a thread in the fabric of a global movement grounded in justice, healing, and renewal.

So take a breath. Celebrate what you've already done. Then ask: What's one thing I can do today from a place of care? Stewardship isn't about carrying the weight of the world; it's about lifting together, with hope and heart.

Celebrate Progress, Not Perfection

Environmental action isn't a checklist to complete; it's a lifelong journey. Along the way, there will be wins and setbacks, days of motivation and days of uncertainty. But what matters most is that you keep going. In a world that often tells us we're not doing enough, choosing to celebrate your progress is a radical act of resilience.

Did you talk to your school about waste? Switch to a plant-based meal once a week? Plant a tree with your family? Those actions ripple outward. Each one is a building block for a more sustainable future. You don't have to do everything. You don't have to do it all at once. But when you do something, and then another thing, and then share what you've learned, you're making change.

Perfection can be paralyzing. It can whisper that if we can't solve it all, we shouldn't bother at all. But the truth is, no one sustains a movement by chasing flawless performance. Movements thrive on participation, persistence, and people who care enough to try again and again.

So pause, look back, and celebrate every step you've taken, every conversation you've started, every shift in habit, every time you chose hope over helplessness. Environmental stewardship is a practice, not a prize. And your progress, no matter how small, is worth celebrating.

A Final Invitation: Be the Example That Ripples Out Into the World

Change doesn't always begin with policy or protest. Sometimes, it begins with one person choosing differently. A neighbour composting for the first time. A student asking where their food comes from. A teacher replacing plastic laminates with reusable materials. These acts may seem small, but they send ripples, visible and invisible, through families, communities, and culture.

You don't have to be an expert or a public figure to make an impact. Just living your values with consistency and kindness is a form of quiet leadership. When others see you reducing waste, speaking up at a meeting, planting native flowers, or organizing a school cleanup, they notice. They wonder. And sometimes, they follow.

You might never know who you've inspired. You might not see all the fruits of your labour. But that's the nature of stewardship: it's about sowing seeds for a future you may not personally harvest. It's about modelling courage and care in a world that often encourages convenience over conscience.

So step forward, not with pressure, but with purpose. Be the example. Your actions have power. Your voice can shape conversations. And your presence, when guided by compassion and conviction, becomes a beacon for others. The road ahead isn't easy, but you never walk it alone.

Thank You, Steward.

If you've made it to the end of this guidebook, thank you, not just for reading, but for caring, for reflecting, for trying, for asking questions, for choosing to be part of the solution. The Earth needs people like you: engaged, thoughtful, hopeful, and willing to act.

We hope this book has sparked new ideas, helped you see the systems around you more clearly, and reminded you of your power. Whether you're just beginning your journey or have walked this path for years, your efforts matter. You matter.

This isn't the end. It's a launch point. There are many chapters still to write, locally, globally, and personally. Keep showing up. Keep connecting. Keep learning. The work of stewardship is not a solo story. It's a chorus, and your voice belongs in it.

Invitation

- **Tell us what you've done.** Share your story, your club, your art, your project. Inspire someone else to begin.
- **Start a conversation.** With a neighbour, a friend, a teacher, a policymaker. Ripple effects grow from dialogue.
- **Pass it on.** Give this book to someone else. Use the worksheets with your students or team. Leave a copy in a library.
- **Stay connected.** Join online communities or networks that energize you. Stewardship is easier and more joyful together.

From all of us who helped bring this guide to life:

Thank you for walking gently, bravely, and wisely forward.

Appendix A

References

References and Sources

This guidebook draws upon a wide array of credible sources from science, Indigenous knowledge, policy research, and grassroots action. Key references include:

Scientific Reports & Global Data

- Intergovernmental Panel on Climate Change (IPCC) Sixth Assessment Report
- UN Environment Programme (UNEP) Global Environment Outlook
- Food and Agriculture Organization (FAO) – State of the World's Forests
- World Wildlife Fund (WWF) – Living Planet Report
- International Union for Conservation of Nature (IUCN) Red List
- The Lancet Planetary Health Journal

Policy and Advocacy Resources

- Project Drawdown
- The Ellen MacArthur Foundation (Circular Economy)
- Environmental Justice Atlas
- UN Sustainable Development Goals (SDGs) Framework
- Global Footprint Network

Community and Case Study Sources

- Zero Waste Europe and Plastic Free Communities
- Indigenous Climate Action (Canada), Aotearoa/New Zealand Environmental Councils
- Fridays for Future, Sunrise Movement, Extinction Rebellion (for youth climate strikes)
- National Geographic, Grist, Yes! Magazine, and Mongabay (environmental journalism)
- Reports and data from local governments and nonprofits cited throughout case studies

Educational Tools & Frameworks

- The Natural Step
- Eco-Schools International
- Earth Charter Initiative
- Leave No Trace Principles
- Open-source curriculum materials from Teach the Future, Earth Rangers, and Green Teacher

URL sources referenced or recommended

(All links are verified as of August 2025)

Carbon Tracking & Climate Action Apps

- Wren: https://www.wren.co
- Earth Hero: https://www.earthhero.org
- Giki Zero: https://zero.giki.earth

Ethical Shopping & Safer Product Tools

- Good On You: https://www.goodonyou.eco
- EWG Healthy Living: https://www.ewg.org/apps
- DoneGood: https://donegood.co

Nature, Advocacy, and Cause-Based Platforms

- EcoHero: https://ecohero.app
- iNaturalist: https://www.inaturalist.org
- Buycott: https://buycott.com

Environmental Education & Policy Resources

- Project Drawdown: https://drawdown.org
- All We Can Save Project: https://www.allwecansave.earth
- Earthjustice: https://earthjustice.org
- UNEP – United Nations Environment Programme: https://www.unep.org

Youth & Community Organizing

- Fridays for Future: https://fridaysforfuture.org
- Sunrise Movement: https://www.sunrisemovement.org
- Our Children's Trust: https://www.ourchildrenstrust.org

Learning Platforms & Tools

- National Geographic Education: https://education.nationalgeographic.org
- WWF Learn: https://www.worldwildlife.org/teaching-resources
- Green School Network: https://www.greenschool.org

Appendix B

Reference Article URLs

1 https://teara.govt.nz/en/kaitiakitanga-guardianship-and-conservation

2 https://vajiramandravi.com/current-affairs/bishnoi-movement/

3 https://riseourworldheritage.org/wp-content/uploads/2024/12/RISE_Nepal_Digital_White_Paper.pdf

4 https://www.undp.org/pacific/stories/learning-traditions-indigenous-fishing-wisdom-solomon-islands

https://marine-conservation.org/on-the-tide/how-fishing-communities-use-locally-managed-marine-areas/

https://www.climatechampions.net/news/samoan-communities-revive-fisheries-with-customary-fishing-rights/

5 https://farmonaut.com/usa/regenerative-agriculture-market-booms-in-north-america-2024

https://www.nature.org/en-us/what-we-do/our-priorities/provide-food-and-water-sustainably/food-and-water-stories/north-america-agriculture/

6 https://earth.org/what-does-carbon-footprint-mean/

https://www.nature.org/en-us/get-involved/how-to-help/carbon-footprint-calculator/

7 https://denmark.dk/innovation-and-design/clean-energy

https://www.jstor.org/stable/24145529

8 https://www.renewable-ei.org/en/activities/column/REupdate/20250508.php

https://www.reuters.com/sustainability/climate-energy/germanys-solar-installations-up-35-early-2024-2024-06-18/

9 https://guidetoiceland.is/best-of-iceland/geothermal-power-in-iceland

https://adventures.is/blog/geothermal-energy-iceland/

10 https://climatecosmos.com/sustainability/how-costa-rica-achieves-99-clean-energy-power/

https://www.iosd.org/harnessing-the-sun-costa-ricas-journey-to-100-renewable-energy/

11 https://www.pembina.org/pub/decarbonizing-remote-indigenous-communities

https://www.microgridknowledge.com/distributed-energy/article/11427583/what-can-we-learn-from-indigenous-communities-about-microgrids

https://nit.com.au/08-05-2025/17831/indigenous-led-energy-projects-drive-investment-and-power-stability-in-remote-communities

12 https://nit.com.au/08-05-2025/17831/indigenous-led-energy-projects-drive-investment-and-power-stability-in-remote-communities

https://livetoplant.com/the-role-of-community-gardens-in-promoting-urban-biodiversity/

13 https://www.oceanicsociety.org/our-work/sea-turtle-conservation/

https://conserveturtles.org/

14 https://www.planetcustodian.com/wildlife-corridors-in-india/21049/

https://www.wildlifeconservationtrust.org/the-curious-case-of-indias-wildlife-corridors/

15 https://www.planetcustodian.com/worlds-largest-urban-wildlife-crossing/24605/

https://www.forwardpathway.us/impact-of-urbanization-on-mountain-lions-and-wildlife-behavior

16 https://www.buglife.org.uk/our-work/b-lines/

17 https://exploringanimals.com/wildlife-corridors-saving-species-from-extinction/

18 https://www.undp.org/rwanda/blog/rwandas-remarkable-journey-inspiring-solutions-global-plastic-pollution-crisis

https://gggi.org/rwanda-lessons-learnt-from-a-pioneer-in-the-fight-against-plastic-pollution/

19 https://wattcrop.com/zero-waste-cities-case-studies-and-future-prospects/

https://www.weforum.org/stories/2019/01/the-inspiring-thing-that-happened-when-a-japanese-village-went-almost-waste-free/

20 https://www.how-to-germany.com/pfand-in-germany/

https://liveingermany.de/pfand-in-germany-bottle-deposits/

21 https://www.no-burn.org/wala-usik-zero-waste-sari-sari-stores/

https://danaasia.org/eco-sari-sari-store-model-bringing-affordable-basic-goods-reducing-plastic-waste/

22 https://aln.africa/insight/turning-the-tide-kenyas-ban-on-plastic-garbage-bags-signals-shift-towards-sustainable-waste-management/

https://www.unep.org/news-and-stories/story/kenya-emerges-leader-fight-against-plastic-pollution

23 https://earth.org/sweden-waste-to-energy/

https://www.energimyndigheten.se/en/news/2025/from-waste-to-wealth---swedens-comprehensive-approach-to-energy-recovery/

24 https://eurochile.cl/en/noticias/the-rep-law-in-chile-promoting-sustainable-waste-management/

https://www.iea.org/policies/16005-law-20920-establishment-of-a-framework-for-waste-management-extended-producer-responsibility-and-recycling

25 https://www.australianbeverages.org/initiatives-advocacy-information/container-deposit-schemes/

https://www.reloopplatform.org/a-brief-look-at-deposit-systems-in-australia-new-zealand/

26 https://www.hrpub.org/download/20240630/CEA32-14834484.pdf

https://www.reuters.com/article/us-jordan-recycling/jordanian-women-go-door-to-door-recycling-trash-idUSKBN1ZQ22P/

27 https://byebyeplasticbags.org/

https://rethinkingplastics.eu/component/content/article?id=188&Itemid=182

28 https://www.palaupledge.com/

https://palau.co/conservation-sustainability/palau-pledge/

29 https://riseourworldheritage.org/wp-content/uploads/2024/12/RISE_Nepal_Digital_White_Paper.pdf

https://pubs.sciepub.com/env/4/3/3/

30   https://discoverwildscience.com/meet-the-forest-guardians-indigenous-tribes-fighting-to-save-the-amazon-1-325517/

https://e360.yale.edu/features/for-the-kayapo-a-long-battle-to-save-their-amazon-homeland

https://www.greenpeace.org/international/story/59198/hyundai-excavators-illegal-mining-amazon-indigenous-lands/

31 https://thorncliffeparkurbanfarmers.ca/

32 https://www.corporate-rebels.com/blog/lessons-from-the-mondragon-cooperative-movement

33   https://www.treesforlifeoregon.org/news/resolution-on-tree-amendments-and-code-overhaul-passes-with-2020-timeframes

https://birdallianceoregon.org/blog/a-nice-win-for-portlands-trees/

34 https://www.plu.edu/scancenter/exhibitions/skolstrejk-for-klimatet-greta-thunberg-and-swedish-school-strikes/

35 https://www.cnn.com/world/live-news/global-climate-strike-students-protest-climate-inaction-intl/

36 https://thegreatgreenwall.org/about-great-green-wall

https://education.nationalgeographic.org/resource/great-green-wall/

37 https://oceanographicmagazine.com/features/exploring-the-coral-triangle-raja-ampat/

https://coral.org/en/where-we-work/coral-triangle/

38 https://otca.org/en/about-us/

https://www.gov.br/mre/en/subjects/international-mechanisms/regional-integration/amazon-cooperation-treaty-organization-acto

39   https://www.environmentagency.no/areas-of-activity/international-cooperation/environmental-cooperation-in-the-arctic/

https://arctic-council.org/news/environmental-protection-an-award-winning-effort/

Appendix C

Acknowledgements

To Indigenous Peoples and Traditional Knowledge Holders:

Thank you for your long-standing guardianship of the Earth. Your wisdom, resistance, and relational worldview are central to this guidebook's values and vision.

To the Scientists and Researchers:

Your tireless work, often behind the scenes, provides the evidence and insight needed to make informed choices and guide effective action.

To the Educators, Artists, and Organizers:

You help translate complex issues into something people can feel, understand, and act on. Thank you for bridging head, heart, and hands.

To the Everyday Stewards and Changemakers:

Whether you're composting in your apartment, organizing your neighbourhood, or fighting for clean air, your actions matter. This book was written with you in mind.

To the Young People Rising:

You are not just the future—you are the fierce, creative, necessary force of the present. Your courage continues to inspire.

Author's Gratitude

This book would not exist without the energy and encouragement of people working for change in every corner of the world—many of whose names will never appear in the headlines.

Thank you for reading. Thank you for caring. Thank you for stepping up.

Let's keep walking this road—together.

Thanks to OpenAI for their invaluable help in the creation of this book.

9781968619503